HOW THE WORLD WORKS
THE PERIODIC TABLE

HOW THE WORLD WORKS
THE PERIODIC TABLE

From Hydrogen to Oganesson

Anne Rooney

ARCTURUS

With thanks to my father, Ron Rooney, FRIC, and Tom Lehrer,
for encounters with elements both benign and dangerous.

ARCTURUS

This edition published in 2019 by Arcturus Publishing Limited
26/27 Bickels Yard, 151–153 Bermondsey Street,
London SE1 3HA

ISBN: 978-1-78888-092-3
AD006354UK

Printed in China

Contents

ORGANIZING PRINCIPLES

'A common chemistry . . . exists throughout the universe.'
William Huggins, pioneer of astronomical spectroscopy, 1909

The world around us is crammed with diverse matter. Things, living and non-living, are made up of millions of different chemicals. Yet all those chemicals are made up of a fairly small number of basic ingredients, known as the 'chemical elements'. Our model for matter is a series of 118 elements, each given their properties by their unique type of atom. These elements combine in myriad ways to make all the chemical compounds that exist in nature or can be manufactured in factories and laboratories.

Astronomer William Huggins recognized that the elements are fundamental: the same elements exist beyond Earth and throughout the universe. The 'common chemistry' he referred to is set out in the Periodic Table of the Chemical Elements – one of the iconic accomplishments of the last 200 years. How the Periodic Table brought order to the chaos of matter around us is one of the greatest stories of science. It helps

Fireworks make spectacular use of chemical reactions; an understanding of the Periodic Table enables chemists to predict the colours that will be produced by different chemical ingredients.

WHAT ARE YOU MADE OF?

Although there are 118 elements, some play a much greater role than others. For example, in your body there is little else besides oxygen, carbon, hydrogen and nitrogen, with some phosphorus and calcium. Oxygen predominates, at 61 per cent (by mass). Then comes carbon (23 per cent), hydrogen (10 per cent), nitrogen (3 per cent), phosphorus (1 per cent) and calcium (1 per cent). There's a smattering of other elements, but only potassium, sulphur, sodium, chlorine and magnesium rate over 0.1 per cent. Even the iron that carries oxygen in your blood accounts for only 0.006 per cent: that's just 60 parts per million (ppm). The human body is a very finely balanced chemical machine, but with relatively few chemical parts.

The seas of the world are almost entirely made up of hydrogen, oxygen, sodium and chlorine, with tiny amounts (less than 0.1 per cent each) of magnesium, calcium, potassium, sulphur and bromine. The Earth's crust has a bit more variety, but even there 98.8 per cent comprises just eight elements: oxygen, silicon, aluminium, iron, calcium, sodium, magnesium and potassium.

us understand how the chemical elements behave and combine, and enables us to predict the course of the chemical processes in which they are involved. Armed with this understanding, we can harness the power of chemistry to make entirely new substances and use it to serve our needs, from curing disease to unleashing the power of the atom.

From chaos to chemistry

The Periodic Table we use today is one of the most information-dense scientific documents ever produced. It organizes the elements in an order dictated by the structure of their atoms. This structure, in turn, determines their properties and behaviour. But no one knew about the structure of the atom when the table was begun. Indeed, no one was even certain that atoms existed at all.

Chemists first attempted to set out the Periodic Table in the 1800s, but its story begins long before that. Considering the nature of matter, the philosophers of Ancient Greece proposed both an early version of atoms and a restricted set of 'roots' of matter that could be combined in different proportions to make everything around us. Since ancient times, the path to the modern Periodic Table has been neither easy nor straight. It veered way off course for more than 2,000 years, and only got back on track around 1660.

The story we trace here centres on three major findings: the realization that matter comprises a limited number of basic chemicals; the discovery of the chemical elements; and the recognition that matter is made of atoms. It takes in the work of chemists, philosophers, hopeful alchemists and nuclear physicists. It meets the victims of terrible accidents and negligent abuse, and people devoted to scientific progress. On the whole it is a story of cooperation, of people working together across time and space to further a single goal: to understand what makes our universe the way it is.

CHAPTER 1

WHAT'S THE MATTER?

'Why not other elements besides fire, air, earth and water? There are four of them, just four, those foster parents of beings! What a pity! Why aren't there forty elements instead, or four hundred, or four thousand? How paltry everything is, how miserly, how wretched! Stingily given, aridly invented, heavily made!'

Guy de Maupassant,
French writer, 1887

What is matter made of? It's a question that has occupied people for millennia. The notion that the myriad types of stuff around us are put together from a limited number of ingredients is age-old. The earliest attempts to answer the question looked for a tiny number of basic components or elements.

The four elements identified by the Ancient Greek philosopher Empedocles were fundamental to the model of the world still embraced in 1472, when this edition of Lucretius' De rerum natura *was produced.*

First things first

Many early cultures had creation myths in which all matter is created from a void, often by a god, as order from chaos. Matter might then be shaped from a single original material or element, such as water or fire, or it may comprise a few elemental types mixed together to give the variety of substances we encounter. Babylonian cosmology had gods that could be considered deifications of earth, water, sky and wind. This seems to have fed into later Egyptian and Greek models, which have four elements, or roots, of matter: earth, water, air and fire. The separation of earth and water, and of day and night or light and darkness, is found in many origin and creation myths, including those of the Abrahamic religions.

It's hardly surprising that earth, water and sky should feature so prominently in early ways of thinking about matter. They represent three domains in which living things are found. And they exemplify the three states of matter: solid, liquid and gas. These states are now recognized by science, but the differences between, say, solid rock, liquid water, and the air we breathe are obvious even outside a scientific framework.

The favourite four (or five)

Throughout much of the ancient world, earth, water, air and fire were considered fundamental elements. The Ancient Greeks may have inherited this belief

> 'And God said, Let the waters under the heaven be gathered together unto one place, and let the dry land appear: and it was so.'
> Genesis, chapter I, verse 9

Egyptian myths of the 3rd and 2nd millennium BC explained creation as a series of births. Here, the air god Shu (centre, kneeling) raises the sky goddess Nut (blue with stars) to separate her from the earth god, Geb (bottom, brown with leaves). The boat carries the sun god Ra, sailing from east to west each day.

from Egypt, and the Ancient Egyptians perhaps derived it from Mesopotamia. The same four elements feature in early Chinese and Indian writings, too. Buddhist texts refer to four and the Hindu Vedas list five elements: earth, water, air, fire and something that could be regarded as void, space or *aether* (see page 18). Another Chinese tradition lists five elements (though this might not be the best term for them): wood, fire, earth, metal (gold, effectively) and water.

The elements were originally regarded as properties or qualities rather than physical components. The Chinese five elements are perhaps better thought of as states of being, or energies. In the Eastern tradition, the 'elements' (for want of a better

word) contributed to a model of medicine, spirituality and being, but did not evolve into an empirical chemical science. For that, we need to turn to Ancient Greece.

Elementary Greeks

The Ancient Greeks were the first to seek to explain the natural, physical world without recourse to the supernatural – the gods – but instead as a chain of physical causes. This model of the world, which we now call the scientific model, seems to have originated with Thales of Miletus, a pre-Socratic philosopher, astronomer and mathematician, who lived in Miletus (now in Turkey), *c*.624–*c*.546BC. Unfortunately, no writings by Thales survive; we have to rely on later accounts of his parascientific thinking. In his major work, *Metaphysics*, the Greek philosopher Aristotle (384– 322BC) recorded and examined Thales' cosmological ideas.

A wet, wet world

Thales maintained that water was the original principle of all matter. To understand this model, we need to accept the Greek distinction between substance and qualities. The substance of everything can be water, but we don't experience everything in the world as water because what we perceive are the qualities of things (matter). It's not such an alien concept. We would now accept that everything comprises atoms and that atoms are all made up of identical subatomic particles,

Earth, water and sky are seen everywhere in the Greek archipelago, a collection of islands in the Mediterranean Sea.

yet we don't experience the particles as such. We experience, for example, the colour and hardness of gold and the fluidity of water. It's a concept found in other early accounts of the world. The Book of Genesis starts with God creating Heaven and Earth and notes that 'the Earth was without form, and void'. The Earth is given substance, but does not yet have form.

'Greek philosophy seems to begin with an absurd notion, with the proposition that water is the primal origin and the womb of all things. Is it really necessary for us to take serious notice of this proposition? It is, and for three reasons. First, because it tells us something about the primal origin of all things; second, because it does so in language devoid of image or fable; and finally, because contained in it, if only embryonically, is the thought, "all things are one".'

Friedrich Nietzsche, *Philosophy in the Tragic Age of the Greeks*, 1873

Other candidates for 'the one'

Thales' model didn't go unchallenged. Anaximenes (*c.585–c.528*BC) thought that air was the primordial element, or *arche*; Heraclitus (*c.535–c.475*BC) plumped for fire. Xenophanes (*c.570–c.475*BC) claimed that everything is made from earth and water. Anaximander (*c.610–c.546*BC), a pupil of Thales, proposed something known as *apeiron*, which means 'infinite' or 'limitless'. He conceived it as an endless mass that yields all that is or will ever be. Anaximander saw the existence of the physical universe as the result of separating out the elements (earth, water, air and fire) from the *apeiron*. When something is utterly destroyed, it returns to the *apeiron* – back to formlessness.

From one to four

Whether we settle for *apeiron*, fire, water or air as the *arche*, we are still left with the need to make varied substances. Difference has to be teased out of uniformity.

Pre-Socratic philosopher Empedocles (*c.490–c.430*BC) resolved all matter into four 'roots' (only later called elements). These were earth, water, air and fire. In his model, all forms of matter are the result of mixing these four roots in the appropriate

The four 'roots' of matter – earth, water, air and fire – and their associated properties came to be associated with the four 'humours' that defined an individual's state of health and age.

proportions. The roots are each associated with two pairs of qualities: earth is cold and dry, water is cold and wet, air is hot and wet, and fire is hot and dry. These four roots and their properties came to underlie all the physical sciences, medicine, and even the precursors of psychology for two millennia.

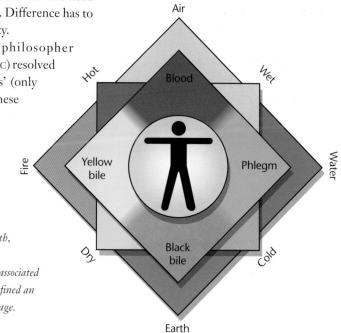

ANAXIMANDER (c.610–546 BC)

Little is known of Anaximander's life except that he was born in Miletus, on the coast of modern-day Turkey. He was a student of Thales and succeeded him as head of the Milesian school of philosophy. He probably taught Anaximenes and Pythagoras.

Anaximander is sometimes considered the first true scientist, and one of the first philosophers to record his ideas in writing, though only one fragment of this work survives. He turned his attention to many different subjects including astronomy, cosmology, geography, meteorology and perhaps biology. He proposed that the Earth hangs unsupported in space, that the Sun is very large (and therefore far distant) and that the celestial bodies are at different distances from the Earth. From the examination of fossils, he concluded that life came first from the sea. He explained rain as a result of the Sun's action on the Earth, creating humidity, and gave an account of thunder and lightning as elemental processes with no divine component.

Above: Considering fossils of marine animals led Anaximander to the conclusion that life originated in the sea.

'Nothing will come of nothing'

Empedocles was clear that matter cannot be created or destroyed – it can only move around. The processes of growth, change, destruction and renewal are all accomplished by swapping around the four roots. In this regard, his model accords with the law of the conservation of mass, demonstrated experimentally and stated by French chemist Antoine Lavoisier in 1774. Empedocles, who lived 2,200 years earlier, was not an experimental scientist. His ideas were rooted in thinking about how the natural world might be organized rather than in designing a testable model to explain observed physical evidence.

The perfect sphere and the two powers

Empedocles' model was part of an entire cosmogony – a theory about the nature and history of the universe. As well as the four roots, he proposed two powers, which he called Love and Strife, that together shape matter through its changing states. The original position of matter has all the four roots in pure and unmixed form occupying a sphere, held together by the attractive power of Love. The edges of the sphere are 'policed' by the repulsive power of Strife which prevents any matter straying beyond the boundaries. Over time, the power of Strife grows and matter can move around, mix and separate out into the forms we see. Over even more time, Love will rise again and eventually all matter will return to its original state. Then the process will begin again. The idea of two forces, one attractive and one repelling, is familiar in modern physics – like electromagnetic charges repel one another and unlike charges attract one another. Gravity draws matter together and dark energy drives it apart. In modern thermodynamics, entropy serves the role of Strife, allowing matter to dissipate and flow in a disordered state.

The four elements, mixed in primordial chaos, depicted by Robert Fludd in 1617.

Something and nothing

There are two possible ways of thinking about the matter in the universe, and the Ancient Greeks suggested both of them. They believed it was either continuous or composed of particles existing in empty space. The second view, called 'atomism', requires there to be not just something (matter) but also nothing (space, or a void) to separate the matter. The concept of empty space has troubled thinkers throughout history.

Starting with atoms

The Greek philosopher Leucippus (dates uncertain) and his pupil Democritus (c.460–c.370BC) expounded the atomist view in the 5th century BC. Leucippus proposed that there is a vast void in which exist invisibly small particles of matter, that he called the 'uncuttables' or *atomos*. We call them 'atoms'. Although these early atoms don't greatly resemble atoms as we know them today, there are similarities. Leucippus thought all atoms were solid (they contain no empty space – a point of division with modern atomic theory); that they were indivisible (we now know there are subatomic particles);

that they are all made of the same ultimate material (in modern theory atoms are all made of identical and interchangeable electrons, neutrons and protons); that they come in different shapes and sizes (atoms of different elements are indeed different sizes); and that everything we see is made up of atoms combining together, and the qualities of matter are produced by the atoms in it.

It's interesting to speculate how the history of chemistry might have developed had the ideas of Leucippus and Democritus prevailed. In the 5th century BC, Leucippus proposed a genuine void – not just an empty space with air in it, but space with nothing in it at all. Unfortunately, the influential philosopher Aristotle thought this untenable, and his views predominated. Without a void there can be no discrete particles because there is nowhere for them to exist where they can be kept separate from other particles. Rejecting a void means accepting

Right: Those Greeks who accepted the concept of atoms assigned one of the original four Platonic solids to each element. Atoms of earth were thought to be the solid cube, of water the slippery icosahedron, of fire the spiky and painful tetrahedron and of air the octahedron. When a new solid, the dodecahedron, was discovered it was assigned to aether.

ATOMS AS ENERGY

Both Hinduism and Buddhism had versions of atoms. In the 7th century AD, a version of atomism flourished in Buddhist India. Led by the philosopher Dharmakirti, it contested that atoms were point-sized and made of energy, with no material form. Physical things were created from this energy, which took the form of qualities. There is a striking parallel between this belief, 1,400 years old, and the modern paradigm of atoms which suggests that they can be described in terms of energy and forces and are almost infinitely small.

a model in which the universe is a block of continuous matter.

So the die was cast: atomism and the void were confined to the intellectual dustbin for the next 2,000 years.

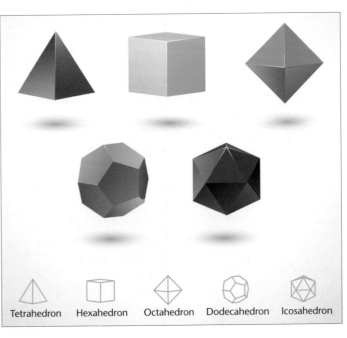

Tetrahedron   Hexahedron   Octahedron   Dodecahedron   Icosahedron

The four elements take off

In Greece, Empedocles' four roots came to dominate thinking about the material world. With these four roots, mixed in different proportions, he argued that it was possible to create the immense diversity of the empirical world, much as a painter can depict that diversity by blending three or four colours in differing proportions. It is a principle that still holds today: changes in matter come about from reordering

Above: Just as an artist can make all hues from red, yellow, blue, white and black paint, all matter could, Empedocles believed, be constructed from just four roots.

and recombining atoms in different formulations, but atoms are not generally destroyed or themselves changed. (There are, of course, exceptions to this general rule, but in the way of everyday life most matter observes it.)

'As painters, men well taught by wisdom in the practice of their art, decorate temple offerings when they take in their hands pigments of various colours, and after fitting them in close combination – more of some and less of others – they produce from them shapes resembling all things, creating trees and men and women, animals and birds and water-nourished fish, and long-lived gods too, highest in honour; so let not error convince you in your mind that there is any other source for the countless perishables that are seen, but know this clearly, since the account you have heard is divinely revealed.'

Empedocles, fragment 23

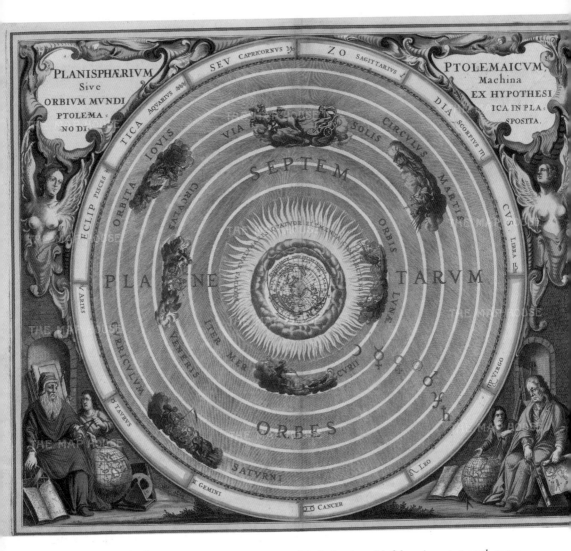

Above and below

The theory of the four roots reflected, or was reflected in, the division of the cosmos into four regions: the earth lay at the centre, with water flowing over its surface, air above it and a realm of fire beyond.

Although Aristotle was content with the four elements for terrestrial purposes, he added another element for the celestial

The Ptolemaic model of the universe puts earth, water, air and fire in the centre, in the terrestrial realm; the other concentric spheres are filled with aether.

realm – the spheres of the heavens beyond the Moon. This was *aether*, an insubstantial, weightless, unchanging substance which filled all space not occupied by other matter. This removed the need for a void, as emptiness was filled with *aether*. Unlike the other elements, *aether* could not combine to form different types of matter, nor did it have properties. It was the element of the crystalline spheres that were thought to encircle the Earth and hold the Moon, Sun, planets and fixed stars. The perfect circular motion of the spheres held the heavenly bodies in orbit around the Earth in a geocentric universe, which was the preferred cosmological model of the time (and until 1543).

Taking it literally

Not all early scientists took the notion of the four elements entirely literally. To borrow from Plato's doctrine of forms, the manifestations of the elements we see around us – the 'earth' that is soil in the garden, the 'water' that flows in a river – are all imperfect renditions of the perfect 'form' of the substances. The perfect, rarified versions are considered the roots of all matter. You couldn't take a handful of soil from the garden and mix it in careful proportions with river water, the flame of a burning fire and the air around you to make a completely different substance, such as mercury or alcohol. Firstly, you would not know the correct proportions but, more importantly, none of the everyday manifestations is pure. You would need to use the *essence* of earth, water, air and fire (though later this type of purification and transformation would become the goal of the alchemists – see pages 21–5).

Nevertheless, we observe that plants take in earth, water, air and the Sun's rays and transform these ingredients into fruit, flowers and leaves, and into wood from which we can make useful objects. And the Earth itself seems to produce the metals, minerals and gemstones that are found within it. So the notion of transmutation is not entirely or immediately ridiculous.

Fast forward

The intellectual legacy of the Ancient Greeks passed to the Romans and then through Turkey, Egypt and Syria into Arab culture. Because they were more engaged in practical progress than in abstract thought, the Romans made no advance on Greek thought in the realm of matter. In the 8th and 9th centuries, the early years of Islam, the Caliphs expended huge effort and expense on centres of learning where scholars were tasked with collecting and translating into Arabic the works of the Ancient Greeks. Thanks to generations of Islamic scholars and their patrons, the works of the Greek philosophers and early scientists have been embedded in European culture. In many fields, Arab scientists built expertly on the works of the Greeks so that when the legacy passed back to Europe by way of translation into Latin from the 11th century onwards, it had been augmented and enhanced by some brilliant minds. In the case of the elements, the original four stood firm. They had, however, become embroiled in the new science of alchemy. Arab alchemists

> 'All nature then, as it exists by itself, has been founded on two things: there are bodies, and there is void in which these bodies are placed, and through which they move about.'
>
> Lucretius, 1st century BC

19

made huge strides in practical chemistry, rigorously exploring all the substances they could get their hands on and developing instruments, compounds and techniques that would be invaluable to later chemists, many of which are still in use today.

Elements of alchemy

While the Greeks studied the composition of matter, they were not experimental scientists. In Ancient Greece, the craftsmen who made potions, dyes, glazes, metals and other chemical products were a world away from the cerebral philosopher-scientists. The two would have seen no points of contact between their realms of expertise; the practical and the theoretical were entirely separate.

Alchemists made the first experimental forays into chemistry. Alchemy was an esoteric science principally concerned with the transformation (or 'transmutation') of matter. Most famously, it led to a 1,500-year search for the 'philosophers' stone', a legendary substance said to be capable of turning base metals into gold.

The fabled Emerald Tablet, shown here as a large document inscribed on a mountain, was supposedly the first, divinely authored, alchemical text.

Origins of alchemy

The Arabic origin of the word 'alchemy' probably derived from *khmi*, the Ancient Egyptians' name for their own country, considered the cradle of chemistry. The earliest Egyptian documents about chemistry, dating from the 3rd century AD, explain how to colour metals to resemble gold or silver, and how to make dyes and artificial gemstones. There is no suggestion that these were attempts to make genuine transformations, just imitations. The myths of alchemy also originate with a character called Hermes Trismegistus, a Greek version of the Egyptian god Thoth. He is said to have ruled Egypt around 1900BC and to have been the author of the 'Emerald Tablet' (*Tabula Smaragdina*), a 'wisdom text' that contained mysterious alchemical secrets. There is no evidence that either the man or the tablet ever existed, but a text purporting to be taken from the tablet first appeared nearly 3,000 years later.

Alchemy flourished under the Egyptians, mainly around Alexandria. We know this because in AD292 the Roman emperor Diocletian demanded the destruction of all 'books written by the Egyptians on the *cheimeia* of silver and gold'. But the occasional act of despotic destruction was never enough to stamp out alchemy entirely, and the knowledge and practice spread with the Arabs, finally reaching Europe in the early Middle Ages. Although intermittently banned by Church and kings, it was pursued enthusiastically well into the 17th century, finally succumbing to the advance of true science in the 18th century. Even the great physicist and mathematician Isaac Newton was a devoted alchemist.

Sometimes described as the 'father of early chemistry', the 8th-century scientist Jabir ibn Hayyan sought to use mercury and sulphur to add properties to 'pure' matter.

Mix and match

The Greek notion that all matter comprises a precise mix of qualities naturally suggests that, if the qualities can be manipulated, one type of matter may be transformed into another. This process lies at the heart of alchemy, which aims to produce undifferentiated matter – a sort of 'ur-matter' (an omnipresent primaeval substance) – and add to it the properties of a desired substance (usually gold).

Alchemists, however, needed to work with something more tangible than principles such as fire and air. Water was acceptable to them, but even a casual observer could see that soil was not the same as some pure form of 'earth'. The answer lay in the work of Jabir ibn Hayyan,

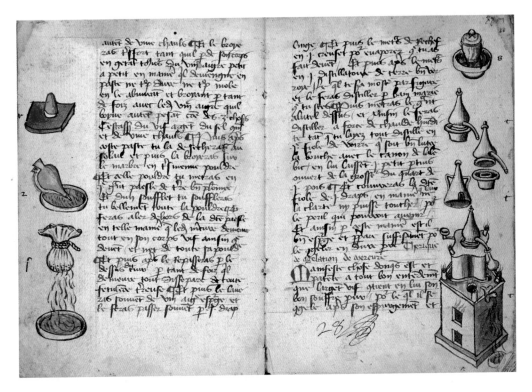

A 15th-century text showing equipment and processes used in alchemy.

thought to have been an 8th-century Arab scientist. Jabir suggested that if the four qualities associated with the elements could be stripped away from matter, it would leave a featureless substance that could be treated as a blank canvas. Other qualities could be added to this, as desired, to make different matter – usually silver or gold.

As a practical scientist, Jabir needed a way of translating these ideas into something that could be achieved in the laboratory. More specifically, he needed to deal with chemical substances, not just ideas of dryness, heat and so on. He suggested that all metals are composed of differing proportions of two naturally occurring substances: sulphur and mercury. In his scheme, mercury provides the wet and cold aspects of matter, and sulphur provides the hot and dry aspects. If the elemental properties could be removed from the starting matter, and mercury and sulphur added in the correct proportions to provide the ratio of hot, wet, dry and cold, then gold would be the result.

A little empirical evidence seemed to support at least one stage of Jabir's theory (although its flaws are clear to us): if the (al) chemist takes water and boils it away, what remains is a white powdery substance that must be pure 'cold'. (In fact the remains would have been whatever impurities had been dissolved in the water at the start – commonly, calcium carbonate, which forms hard-water limescale.)

Jabir stated that only pure mercury and sulphur could be used to make gold – he argued that any impurities would produce other metals. Hopeful alchemists spent centuries trying to perfect their methods of refining and blending their ingredients to achieve the purity they needed. Needless to say, they didn't succeed.

Another Arab alchemist, Muhammad ibn Zakariyya al-Razi (854–925), concluded that many metals contain a salt of some kind. This notion resurfaced in the work of the Swiss chemist and physician Paracelsus (Theophrastus von Hohenheim, 1493–1541) who believed that earth generated all living things under the rule of three 'principles' – mercury, sulphur and salt. These principles, and mixtures of them, he considered to be very potent in chemistry, medicine and toxicology.

Alchemical magic and secrecy

Although it is frequently portrayed as akin to magic, alchemy was logical within the knowledge structures of its time. Even so, it was surrounded by secrecy,

Left: The 9th-century Arab chemist Muhammad al-Razi.

especially once it reached Europe. Partly to protect their commercial interests, as well as for reasons of safety, alchemists jealously guarded their esoteric knowledge. Until the 13th century, non-secular societies tended to regard the alchemists' craft as sacrilegious, and often outlawed it. Alchemists, therefore, were at risk of arrest and persecution.

Below: Cryptic allegorical illustrations and narratives hid the secrets of alchemy from untutored readers.

23

Secrecy extended to alchemical texts being couched in obscure allegories. In the way that the jargon of certain modern disciplines serves to exclude the lay person from the debate, alchemical jargon preserved alchemy for an elite, and increased its aura of mystery.

Many alchemical rituals now look odd, far-fetched or futile, such as deliberating about which phase of the moon is most favourable for collecting ingredients and

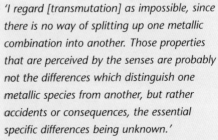

'I regard [transmutation] as impossible, since there is no way of splitting up one metallic combination into another. Those properties that are perceived by the senses are probably not the differences which distinguish one metallic species from another, but rather accidents or consequences, the essential specific differences being unknown.'

Ibn Sina, 11th-century philosopher

carrying out a procedure. These rituals were not contrived simply to make alchemy inaccessible to the uninitiated. Medieval and early modern alchemists inhabited a world in which astrology was taken entirely seriously. The movement and alignment of the heavenly bodies was assumed to have an impact on events on Earth and in human lives. There was the added bonus, of course, that if alchemical procedures couldn't be carried out because the conditions were not ideal, then failure could be explained away without having to challenge the underlying theory. If, for example, the alchemist hadn't been able to collect the correct sample from between the toes of a basilisk during the third phase of the moon, he wouldn't have had to acknowledge the method was invalid – he could excuse its failure.

Two true elements

Although there was still no concept of elements in the current understanding of

A three-headed monster in an alchemical flask represents the composition of the alchemical philosopher's stone: salt, sulphur and mercury. From Salomon Trismosin, Splendor solis, *1530s.*

The flood of 'white water' (mercury) from the base of an oak tree eludes the people digging for the spring in the garden because they are blind – a fair allegorical representation of the failure through ignorance of alchemists to find what they seek.

the term, the two ingredients the alchemists had settled on, mercury and sulphur, are in fact elements. They are not, however, components of the other metals, many of which are also elements, including the silver and gold that the alchemists hoped to make.

But mercury and sulphur were not the only actual elements known to the ancients. In the next chapter we will examine the elements that were known and used, although their status was not yet recognized at that point in history.

CHAPTER 2

ELEMENTS IN PLAIN SIGHT

'Next to copper we must give an account of the metal known as iron, at the same time the most useful and the most fatal instrument in the hand of mankind.'

Pliny the Elder,
Natural History

Everything around us is made up of elements, but that doesn't mean the elements are clearly recognizable. Most substances are compounds – they are made up of more than one element. And most objects and substances we encounter in the world contain more than one type of chemical mixed together. Even so, our early ancestors were familiar with and made extensive use of some of the elements. By putting them to use they explored their properties, and in doing so acquired the first seeds of chemical knowledge.

European cave art, such as this example from Chauvet-Pont d'Arc in France, was frequently made with carbon-based black pigments. The drawing was created around 36,000 years ago.

Seeing what we have

When our distant ancestors learned about the types of matter around them, it was with primarily utilitarian intentions. Which metals would make a good weapon? Which kinds of mineral would make an attractive glaze on a pot? What medium could be moulded easily to make a crown or bracelet, and what could be used to line a water cistern to keep it from leaking?

Elements make history

Some elements exist naturally in their native form and were evident in prehistoric times. More than 7,000 years ago, our ancient

Prized for its deep red-orange colour, ground cinnabar (mercury sulphide) has been used as a pigment in Chinese lacquer-work since 3000BC.

ELEMENTS, COMPOUNDS, MIXTURES

An element, as modern chemists understand it, is a substance in which all the atoms are of a single type – one that cannot be further broken down. There are 118 chemical elements, including such widely differing substances as oxygen, gold, iodine, mercury and carbon.

A compound is a molecular substance in which atoms of two or more types are chemically bonded. The atoms cannot be easily split apart by physical means, though often they can be persuaded to separate and perhaps combine with other atoms to make molecules of one or more different compounds in a chemical reaction. A compound has its own properties and behaviours that may be unrelated to those of its constituent elements. An example is sodium chloride, common table salt, made of the two elements sodium and chlorine, yet resembling neither of them.

A mixture is just as it sounds: a mix of two or more different substances. They are not chemically bonded together and can often be separated by physical means. They can certainly be separated by chemical means. The characteristics of the components remain the same; the mixture does not have new properties of its own as a compound does. An example of a mixture is sand and sugar; the substances can easily be separated by adding water to the mixture, dissolving the sugar and filtering out the sand. Then the sugar can be recovered by evaporating the water. The two original components are unaltered by having been mixed.

ancestors were aware of five metals: copper, lead, gold, silver and iron. Mercury has been known in its native form since around 1500BC. Carbon was widely found in forms of charcoal and soot.

Copper, star of the Bronze Age

Copper was probably first used in the Middle East around 11,000 years ago. It was the first metal to be worked by our ancestors. The mining of native copper began at Çatalhöyük, Anatolia (Turkey) in the late 8th millennium BC. The oldest surviving manufactured metal objects are some copper beads from Anatolia.

Copper was important not just in its native form but as the primary material of the Bronze Age, when metal tools took over from stone implements. Bronze is not an element but an alloy of copper with another element, most commonly tin or the metalloid (semi-metal) arsenic. The Bronze Age started at different times around the world, beginning about 3500BC in Mesopotamia, between 3000 and 2500BC in Western Europe, and around 2000BC in Eastern Asia. The new age was also characterized by the development of writing; the creation of civic centres with legal systems and government; the realization of ambitious architectural projects; the commencement of year-round agriculture; discoveries in medicine; the organization of religion; and the seeds of intellectual endeavour in fields such as astronomy and mathematics. These achievements signalled the dawn of civilization.

Copper can be dug from the ground directly in some places, and has been mined for thousands of years.

Bronze is an alloy of copper and another element, commonly arsenic or tin. As arsenic is a frequent contaminant of naturally occurring copper 'ore' (mineral-bearing rock), the first bronze would have been discovered rather than made. The earliest known copper smelting was carried out in Mesopotamia; the oldest objects made from this process were found at Tepe Yahya, in Iran. This was a significant moment in human history. It saw people living in the 4th millennium BC recognizing and digging up ore and heating it to get at the content they wanted.

Copper ores in the area contain arsenic and it seems that people began deliberately seeking out these ores because the metal produced from them was superior to the native copper. As little as 0.5–2 per cent of arsenic by weight improves the tensile strength and hardness of copper by 10–30 per cent. Where copper occurs near the surface, it usually has a lower arsenic content. Once this resource had been used up, miners dug deeper and found that the metal smelted from this ore, with arsenic, was of better quality. It's likely that the deliberate addition of arsenic-bearing ore followed quite rapidly to make bronze in preference to copper. The manufacture of bronze was one of the most significant steps in human history.

From plumbing to poisoning

The second elemental metal to be worked was lead, probably around 9,000 years ago. Beads made from lead dating from 6500BC have been found in Çatalhöyük, Turkey. Lead has been mined for at least 6,000 years. A soft, heavy metal with a low melting point (327°C), it is easy to work. Although it quickly acquires a thin film of oxide in air (which explains why it's not shiny), lead doesn't corrode and is unaffected by water. This has made it useful in plumbing, an

SMELTING: METAL FROM ORE

The process of smelting is fairly straightforward. Ore containing the metal is heated and the metal reacts with oxygen in the atmosphere to form a 'calx' (oxide). Heating the oxide in an oxygen-poor atmosphere produces carbon monoxide (as there is too little oxygen to produce carbon dioxide directly), and the oxygen-hungry monoxide strips oxygen from the calx, leaving the pure metal behind. Heating rocks in a closed environment was a fairly easy process for early metalworkers to carry out. It's likely that copper and lead were the first metals smelted, during the 6th millennium BC. Copper smelting was practised by the Vinča culture in Serbia and lead smelting might have started in Yarim Tepe (now northern Iraq).

Above: Lead is so durable that Roman plumbing is still serviceable after nearly 2,000 years.

application that probably started with the Ancient Egyptians. It was used most famously by the Romans to make lead piping and cisterns. The word 'plumbing' comes from the Latin for lead, *plumbum*. This is the origin of the chemical symbol for lead, Pb.

Lead was usually produced by burning the mineral galena (lead (II) sulphide). At least 4,000 years ago, the Assyrians used lead currency. A lead statuette from Abydos in Egypt dates from around 3800BC. Lead was also used to make slingshot in Ancient Greece, and lead compounds were added to colour glazes and cosmetics.

Sweet but deadly

Their empire might have lasted longer if the Romans had confined their use of lead to plumbing. But unfortunately they also used a compound of lead in their food. Sugar of lead, or lead acetate, has a sweet taste and became a popular sweetener in Ancient Rome, where it was added to wine as well as food. Roman winemakers commonly boiled unfermented grape juice for long periods to produce a concentrate called *sapa*, which they added to wine to sweeten it. They found

31

that boiling it in lead vessels produced an even sweeter taste than that of juice boiled in copper vessels. The reason for this was that the lead was being leeched from the pan by the action of acetic acid in the grape juice to produce a compound called lead acetate. The Romans soon discovered how to make the sweetener directly by mixing lead oxide with vinegar (acetic acid). Modern chemists reproducing the *sapa* found it contained up to 1g (.035 oz) of lead per litre of grape juice. A single teaspoonful could have caused chronic lead poisoning.

The sweetener was popular and widely used – of 450 recipes in the Roman

Wine drinking in Ancient Rome may have had more lasting health consequences than just a hangover.

cookbook *Apicius*, around one fifth use sugar of lead as an ingredient. Many wealthy Romans suffered from gout, to such a degree that it was commonly satirized as a disease of the self-indulgent and dissolute. It's likely that at least some of these cases were so-called 'saturnine' gout, produced by lead poisoning. Other characteristics of lead poisoning include a low sperm count; infertility was widespread among the later Roman emperors, despite their prodigious attempts at insemination. From AD15–225, many Roman rulers were dull-witted, cruel, and physically or mentally impaired; some historians of medicine regard this as strong evidence of lead poisoning.

Treasures galore

Like lead, gold is soft and easily worked. It doesn't corrode, is undimmed by exposure to air (as it doesn't form an oxide), and is widespread in its native form as seams running through rocks and as nuggets in rivers. Gold seems to have first been used in Ancient Egypt, around 5000BC, in artefacts made of electrum, a naturally occurring gold-silver alloy. Pure gold jewellery was made in Ancient Sumer from 3000BC using gold found as dust, flakes or nuggets and melted down. The Ancient Egyptians mined gold from 2000BC. Although the earliest known use was in the Near East, gold is found around the world and has been used by all cultures. Its chemical symbol, Au, comes from the Latin word *aurum*, meaning 'shining dawn'.

Silver is also found around the world and was exploited by many early societies. It occurs in ores with other metals, most usually lead, and is easily extracted. For the ancients, even ores giving a relatively low yield (1 per cent) of silver could profitably be smelted to provide the metal, especially if the other product – lead, for example – was also useful. Silver is relatively easy to work and has an attractive sheen when polished, making it well worth the effort of extraction for early metalworkers.

Serious mining of silver began around 3000BC. In Ancient Egypt, silver was for a time more valuable than gold as it was more difficult to find, not occurring readily in its native state. The easy extraction of silver was achieved with the invention of cupellation, around 2500BC. This involves burning ore (such as the silver-containing galena) and absorbing one component (such as lead) by its reaction with another substance, to

A plaque of a stag, made in the 4th century BC in Scythia from electrum, a mix of gold and silver.

Silver mining in Kutná Hora, Bohemia (now in the Czech Republic), depicted in the 1490s. Silver was first found here in the 10th century.

isolate the precious metal. The first use of cupellation was in Syria and Turkey in the 4th millennium BC, using a form of galena containing iron and silver. The chemical symbol for silver is Ag, from the Latin *argentum*, meaning 'shiny' or 'white'.

Iron men in an Iron Age

Iron or, more properly, steel, marked a new departure for humankind. Pure iron is susceptible to rust, making it less enduring than many other metals. This means that ancient iron artefacts are relatively rare.

The earliest use of iron was by the Ancient Egyptians more than 5,000 years ago, long before iron smelting had been achieved. The iron they used was meteoric iron (identifiable by its high nickel content). Meteoric iron, or 'metal from heaven', was revered. Because of the nickel it was soft enough to work fairly easily. A dagger made from meteoric iron was found among the grave goods of Tutankhamun.

Iron was first smelted in Syria and Mesopotamia between 3000BC and 2700BC. It was smelted extensively in Anatolia by the Hittites sometime between 1500BC and 1200BC and became widespread after the fall of their empire in 1180BC, at the

---

SILVER AND THE RISE OF ATHENS

Ancient Greece was the cradle of the intellectual burgeoning of modern science, philosophy and literature. Yet none of it might have prevailed without the silver mines of Laurion (or Laurium). Situated south of Athens and first worked around 3200BC, the mines contributed significantly to the city's wealth, funding wars against the Persians in 480BC and 490BC which secured Athenian dominance and enabled the flourishing of Greek culture.

A 7-metre-high iron column in Delhi was erected around AD400 by King Chandragupta II. The remarkable lack of corrosion is the result of the high phosphate content of the iron, which has allowed the formation of a protective layer of hydrogen phosphate crystals on the surface.

'It is by the aid of iron that we construct houses, cleave rocks, and perform so many other useful offices of life. But it is with iron also that wars, murders, and robberies are effected, and this, not only hand to hand, but from a distance even, by the aid of weapons and winged weapons now launched from engines, now hurled by the human arm, and now furnished with feathery wings. This last I regard as the most criminal artifice that has been devised by the human mind; for, as if to bring death upon many with still greater rapidity, we have given wings to iron and taught it to fly.'

Pliny the Elder, *Natural History*, Book 34

end of the Bronze Age and the beginning of the Iron Age. Smelting appeared in Europe about 200 years later. In India, it began perhaps as early as 1800BC.

The magical metal of immortality (or not)

While all the metals discussed so far were put to practical use, one seems at first glance to be entirely useless. You can't forge it into any kind of object, it can poison you, and even extracting it is perilous – but it is beguilingly lovely. Mercury, or quicksilver, is the only metal that's a liquid at room temperature. Once seen, it is never forgotten. Beads of mercury roll around on a surface, then instantly clump together on meeting. They are beautiful to watch – but potentially deadly to play with. Because of this rapid rolling movement, mercury was also known as 'quicksilver'.

Mercury's first use by humans was in naturally occurring compounds; the mercury itself was not apparent. Thirty thousand years ago, Paleolithic people used the mineral cinnabar to produce the bright red pigment vermilion to paint bison and other creatures on the walls of caves in France and Spain. Occasional droplets of pure mercury are found in cinnabar, which is the mineral mercuric sulphide (HgS). Mercury was recovered by heating the cinnabar in air and condensing the mercury vapour that emanated from it – a dangerous practice, to say the least, as mercury is highly toxic as a gas. The chemical symbol for mercury, Hg, comes from the Latin word *hydrargyrum*, meaning 'water-silver'.

Perhaps on account of its beauty and uniqueness, mercury was considered to have

Cinnabar has been used as a pigment in cave paintings since around 25,000BC.

MOATS OF MERCURY

The tomb of Emperor Qin has never been excavated. Located in Xian, China, alongside a necropolis famous for the thousands of terracotta soldiers buried there, the main tomb complex is said to be an entire underground city surrounded by moats of mercury. The mercury content of the soil in the area is certainly very high. The tomb is supposedly cursed and booby-trapped, which has deterred people from excavating it.

extraordinary and (unfortunately) health-giving properties. It was known in China by 2000BC, and has been found in Egyptian tombs in Kurna dating from 1500–1600BC. Qin, the first emperor of China, died in 210BC at the age of 39, apparently from mercury poisoning. He had consumed a large quantity of mercury pills, which he believed would bestow immortality.

Mercury and gold

Aside from its strange appearance, Mercury fascinated the ancients for another reason: it dissolved gold. Mercury was used from early times to extract alluvial gold – gold dust in the sludge found at the bottom of rivers. Gold (and silver) would both dissolve in the mercury while the other contents of the sludge would not. Skimming off the contaminants and roasting the amalgam to evaporate the mercury left the precious metals behind. But it also polluted the environment and poisoned miners. The Phoenicians and Carthaginians used mercury from mines in Spain to recover gold 2,700 years ago. The Roman natural historian Pliny left an account of the process in the 1st century AD, and the Romans expanded gold recovery by this method. Around AD77 they were importing 5,000kg (11,023 lb) of mercury a year from Spain to use in gold amalgamation. But while the Romans seem to have remained ignorant of the perils of lead, they may have made a connection between mercury and environmental health issues as the process was banned within 100 years.

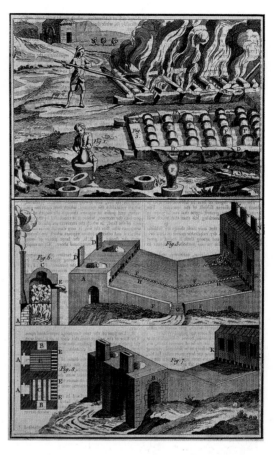

The processing of Mercury to illustrate sublimation, from The Royal & Universal Dictionary of Arts & Science, *1791.*

The use of mercury to extract gold was introduced to South America in 1554. Between 1580 and 1900, an average of 612 tonnes of mercury was used every year in Spanish America. Between 1850 and 1900, 1,360 tonnes a year was used in North America, mostly for gold and silver extraction. During this period around a quarter of a million tonnes of mercury was lost to the environment. It was deployed in the California Gold Rush and into the late 20th century. The widespread use of mercury in gold extraction is thought to be the reason for exceptionally high levels of the metal in the soil of some areas of North and South America (up to 0.5mg per gram in some hotspots). Mercury poisoning from gold mining has recently been reported in the Brazilian Amazon, the Dixing region of China, and the Philippines.

A jar for storing mercury pills, a popular but toxic medicine for centuries.

Health-giving or deadly?

Mercury has had many uses over the centuries, often with unfortunate consequences. Medieval Arab physicians included it in lotions to treat skin disease. In the 10th century, the physician and alchemist al-Razi experimented with its toxicity on animals. A hundred years later, the physician Ibn Sina wisely decided that mercury should be used for external treatment only.

Maverick Swiss physician Paracelsus had no such scruples, and promoted its use as a medicine in medieval Europe. The sexually transmitted disease syphilis emerged in Europe in the late 15th century, and in 1530 Paracelsus wrote that mercury was an effective treatment (it kills the coil-shaped bacteria, called spirochetes, that cause the disease). The treatments for syphilis which developed over the coming centuries included fumigation with mercury vapour; swallowing tablets containing mercury; and using lotions or potions laced with mercuric chloride (called 'calomel'). The side effect – mercury poisoning – was deeply unpleasant. But so was syphilis, and that was more likely to be fatal so it was worth the gamble. Unfortunate patients suffered ulceration of the skin, mouth and throat, tooth loss, neurological damage and, sometimes, death. Less serious but still unpleasant were sweating and drooling, which gave the treatment its colloquial name of 'salivation'. Curiously, it doesn't seem to have occurred to people that something marketed as an effective rat poison was also likely to be detrimental to health when ingested as a medicine. Sales of mercuric chloride, especially when mixed with brandy as 'Swieten's liquor', were robust.

Hat-makers who used mercury in the preparation of felt developed a condition

'An effectual remedy for all scorbutic and herpetic eruptions of the face and skin, from the most trivial to the most disfiguring and inveterate; from the smallest pimple or tetter [blister] to the most universally spreading eruptions or ulcerations. For redness of the nose, arms, or other part, and in short for every train and species of evil to which the skin is liable, whether vivid and inflamed, or languid and obdurate.'

The marketing blurb for Gowland's lotion, developed in the mid-18th century, which contained mercuric chloride and/or white lead called 'mad hatter disease'. Symptoms included mental confusion, pathological shyness, irritability, depression, apathy, a terror of being ridiculed, an explosive temper when criticized and, in extreme cases, memory loss, personality change and delirium. The prevalence of the condition led to the epithet 'as mad as a hatter'.

Elements in hiding

Around 1000BC in India miners were extracting zinc, but they were not aware that it was a distinct metal. They used zinc-containing ores to make brass. Zinc was known to the Romans, but they rarely

Although it's tempting to assume that the Mad Hatter in Lewis Carroll's Alice in Wonderland *suffered from mercury poisoning, his gregarious behaviour does not support the idea.*

used it. It wasn't recognized as a discrete metal until 1746, when it was discovered by the German chemist Andreas Marggraf.

The ancients also used antimony and arsenic without realizing that they were distinct from other substances they knew (both are metalloids).

Antimony isn't lead

Antimony was known to the Sumerians in Mesopotamia 5,000 years ago. A fragment of a vessel in the Louvre, in Paris, taken from the ruins of the Sumerian city Telloh, was identified in 1887 by French chemist Marcellin Berthelot as being made of pure antimony. (Repeat analysis in 1975 found it to be only 95 per cent antimony.) Both the 1st-century Greek physician Dioscorides and the Roman writer Pliny refer to the extraction of metal by burning

THE MAD TEA - PARTY.

the ore stibnite (antimony sulphide). They warned that it turned to lead, so apparently were unaware that antimony is a distinct substance.

Antimony compounds were also present in kohl, a cosmetic used to darken the eyebrows and draw around the eyes. King Nebuchadnezzar II of Babylon (604–561BC) had his palace walls built from bricks glazed with a lovely yellow colour derived from a compound of lead and antimony.

Around AD800, Arab alchemists isolated and identified antimony as a separate substance. As a metalloid, antimony was considered to be hermaphrodite (neither male nor female, but partly both). Its alchemical symbol, inverted, is now the sign for 'female'. Antimony is poisonous, but has been used in compounds as an astringent and an ointment for burns and ulcers. As antimony pills upset the bowel and prompted it to empty, they were swallowed as laxatives. The antimony, being valuable and hard to come by, was recovered, washed

---

DID MOZART DIE FROM ANTIMONY POISONING?

The composer Wolfgang Amadeus Mozart died in 1791 at the age of 35 after coming down with a mysterious illness. He was treated with a 'tartar emetic', as antimony potassium tartrate was commonly called. Mozart's subsequent symptoms of severe vomiting, fever and swollen limbs and abdomen are consistent with antimony poisoning. He died two weeks later. The cause of his death was given as 'miliary fever', but this was by no means conclusive.

---

and kept for re-use. There are reports of antimony laxatives being passed down through families.

Deadly, and not so deadly

While arsenic-bearing ores were added to copper to make bronze, it's unlikely that early smelters isolated arsenic as doing so

There has been conjecture that Nebuchadnezzar II, who famously went mad and lived as a cow for seven years, was poisoned by his décor. However, as the antimony glaze would have been sealed to the tiles it's unlikely that this would have been the cause. William Blake's illustration dates from around 1800.

would have been very dangerous. Arsenic sublimates (passes directly from a solid to a gas), so anyone burning arsenic-bearing ores in a closed space is at serious risk. To produce arsenic, the intrepid smelter would need to burn an arsenic-bearing ore and condense the arsenic vapour; a successful outcome would have been unlikely.

One of arsenic's prime uses throughout history has been as a poison – against rats,

The mummified body of Ötzi, a man who died around 3100BC, contains high concentrations of arsenic, leading some archaeologists to speculate that in life he was a copper smelter.

mice, flies and other pests. In Renaissance Italy and France, the compound arsenic trioxide (As_2O_3) was known as 'succession powder' as it was commonly used to dispatch relatives occupying hereditary

THE MARSH TEST

The Marsh Test is named after James Marsh, a chemist at the Royal Arsenal in Woolwich, London, who prepared the scientific evidence against a man accused of killing his grandfather by lacing his coffee with arsenic. The test for arsenic available at the time gave an unconvincing result and the defendant was acquitted. (He later admitted his crime.) Marsh was so frustrated that in 1836 he developed his own test. A sample of tissue or body fluid containing arsenic was mixed with zinc and acid to produce arsine gas. When ignited, the heated gas stained a ceramic bowl with a silvery-black deposit of arsenic. The test proved infallible, detecting as little as 0.02 mg of arsenic, and removing a foolproof weapon from the poisoner's armoury.

MISS MOLLY BLANDY.
who with her own & her Sweethearts Contrivance, did Barbarously and Inhumanly Poison her own Father for his Estate.

office. During the 19th and early 20th centuries, arsenic was the poison of choice for murderers as its symptoms are similar to those of a large number of naturally occurring conditions. Its presence was not detected until the development of the Marsh Test in 1836 (see box on page 41).

Arsenic has also been used as a medicine, both in ancient times and more recently. Paul

Arsenic was a popular choice with murderers. Mary (Molly) Blandy was convicted of killing her father with small doses of arsenic. She was hanged at Oxford on 6 April 1752. In her defence, she claimed she thought it was a love potion which would lead him to accept the man she wanted to marry.

Ehrlich's 'magic bullet' remedy for syphilis, Salvarsan, is an arsenic compound, and is now used to treat leukaemia. In small quantities arsenic has a tonic effect. It has even been used by unscrupulous race-horse owners to improve their animals' performance.

Arsenic was probably first isolated either by Arab alchemists (traditionally, Jabir) or, according to European tradition, by German scientist/philosopher Albertus Magnus around 1250. Arsenic trioxide, heated in vegetable oil, yielded metallic arsenic which could be used to colour copper to look like silver – a useful trick for a dishonest alchemist to know.

White gold is not silver

Platinum was used in the West as early as 700BC, but it was probably not distinguished from silver so is not generally considered to have been an element known to the ancients. The French chemist Marcellin Berthelot, who identified platinum in the so-called Casket of Thebes in 1901, suspected it had not been used deliberately. The platinum decoration on the casket was found to be mixed with small quantities of gold and iridium, and had probably occurred naturally in this form in ores imported from Nubia as a source of gold. Pure platinum only came to light in Europe through items imported from South America by Spanish Conquistadors (see page 84).

Fiery elements

Carbon and sulphur, two elements with close links to fire, were also known to the ancients. Carbon was familiar in the forms of charcoal and soot produced by the burning of wood. It also occurred naturally as graphite, anthracite (coal) and diamond, though it would be a long time before anyone recognized that these disparate substances were all forms of the same material. Carbon in the form of charcoal was used for drawing outlines in Paleolithic cave art (see pages 26–7). Its first industrial use was by the Egyptians and Sumerians, around 3750BC, to reduce copper, zinc and tin ores when making bronze. Diamonds were first known in India between 6000BC and 3000BC and in China around 2500BC.

ARSENIC – YUM YUM!

Consistent exposure to arsenic builds up the body's tolerance to it. The so-called arsenic eating peasants of the Styrian Alps in Austria used to consume 250 mg (0.0088 oz) of arsenic trioxide twice weekly, claiming it improved their skin and enabled them to work better at high altitudes. This dose is more than enough to kill an adult, yet one of the peasants, tested in 1875, had consumed 400 mg (0.014 oz) with no ill effects.

A CEREMONIAL AXE TO GRIND

In 2005, Harvard physicist Peter Lu found that Chinese craftsmen used diamonds to polish ceremonial axes 4,500 years ago. Four axes with highly polished surfaces, and made of the second hardest mineral known, could only have gained their reflective sheen from being buffed by diamonds. The axes are made of corundum, which in its red form is ruby and in its blue form is sapphire. Nothing but diamond is harder than corundum. Lu's study involved polishing a small sample from one of the axes with both quartz (previously assumed to have been used by the ancient axe-workers) and diamond. Close examination with an electron microscope and X-ray diffraction revealed that the surfaces polished with diamond bore the closest resemblance to the surface of the ancient axes. Such a high level of smoothness could not be achieved using quartz.

Burning diamonds

In 1772, French chemist Antoine-Laurent de Lavoisier used a lens to focus sunlight on to a diamond in an atmosphere of pure oxygen. He found that the only product of this burning was carbon dioxide. English chemist Smithson Tennant repeated the experiment in 1796, this time comparing the results of burning the same mass of graphite. Finding both gave exactly the same amount of the same product, he concluded that diamond is simply a form of carbon. This magical vanishing trick had first been demonstrated by two Italian scientists, naturalist Giuseppe Averani and medic Cipriano Targioni, in 1694. They used a magnifying lens to focus sunlight on to a diamond and showed that it disappeared, leaving no trace. Since this was before the identification of either oxygen or carbon dioxide, it yielded no useful conclusion – it was just an expensive party trick.

From under the volcano

Sulphur was known in the ancient Middle East and is mentioned several times in the Old Testament. Most famously it was supposedly the agent of the destruction of Sodom and Gomorrah. The Ancient Greeks burned it as a fumigant to rid their homes of pests. In Homer's *Odyssey*, Odysseus commands his nurse to bring sulphur so that he can purify his house. Mount Etna was a source of sulphur for the Romans, who burned it to obtain sulphur dioxide which they used to bleach cloth and preserve wine.

The alchemists thought that sulphur was one of the elements which make up all metals, so it rose to prominence in the Middle Ages and remained important until the 1700s. Later, Lavoisier claimed it as an element, but his assertion was contested by Humphry Davy, who was convinced that sulphur was a compound containing hydrogen. Davy was proved wrong by French chemists Gay-Lussac and Louis-Jacques Thénard in 1809.

Known, but not recognized

By 1250 at the latest, seven metal elements, two metalloids (arsenic and antimony) and two non-metals (carbon and sulphur) were known and used. But they were not recognized as elements or as pure substances. Before recognition could come, people needed to know rather more about the nature of matter. They would discover this by looking at the invisible – the gases around us.

Mount Etna, in Sicily, is part of a volcanic area that spreads up into mainland Italy. Volcanoes were a source of sulphur to the Ancient Romans, an element they put to good use in incendiary weapons and fireworks.

GREEK FIRE?

Sulphur is implicated in a mysterious liquid known as Greek fire. Used as a weapon by the Byzantine navy, its ability to set a ship ablaze when launched from a distance struck terror into the hearts of enemy sailors. Greek fire was last used in 1453, at a battle that saw the fall of Constantinople (now Istanbul) and the end of the Byzantine Empire. It burned so fiercely that it was impossible to put out – even continuing to burn as it lay on the surface of the sea. The recipe was a fiercely guarded secret and anyone betraying it was put to death. The ingredients of Greek fire are still unknown, but it was probably a mix of stibnite (antimony sulphide), saltpetre (potassium nitrate) and crude oil.

45

CHAPTER 3

A BREATH OF FRESH AIRS

'The generality of men are so accustomed to judge of things by their senses that, because the air is invisible they ascribe but little to it, and think it but one remove from nothing.'

Robert Boyle,
Memoirs for a General History of the Air, 1692
(published posthumously)

The elements put to use by the ancients and explored by the alchemists and Arab scientists were all tangible substances. Air largely escaped their notice. But air would ultimately supply the vital clue to the atomic nature of matter. Work with gases proved beyond reasonable doubt that matter is divided into tiny particles. This, in turn, formed the basis of the modern paradigm of the chemical elements from which the Periodic Table would emerge.

Eventually, understanding the nature of gases would enable their use in hot-air balloons and blimps (small airships), as in this polar expedition planned in 1877.

Out of the void

The Ancient Greeks proposed two alternative models of matter, one of continuous matter and the other of discrete particles existing in a void. The former prevailed virtually unchallenged until the 17th century, when the notion that things might be made of tiny bodies, or 'corpuscles', began to gain ground. The fortunes of the atom began to change in France, with the work of two philosophers, René Descartes (1596–1650) and Pierre Gassendi (1592–1655).

A packed universe

Descartes didn't see the need to choose between the two models inherited from antiquity, even though they seem incompatible. On the one hand Aristotle had denied there could be a void; he said matter was continuous and infinitely divisible. On the other hand atomists such as Leucippus and Democritus had maintained that there is a limit to the size of a particle of matter (an atom is, by definition, an indivisible particle) and that these particles move around in a void. Descartes' model was of a universe packed with matter yet capable of change and movement. This meant, essentially, that matter and space were equivalent. In his scheme, matter was continuous (there was no void), yet minute particles also existed. For Descartes, when something became rarified (that is, when the particles moved further apart), other smaller particles moved to fill the spaces in between.

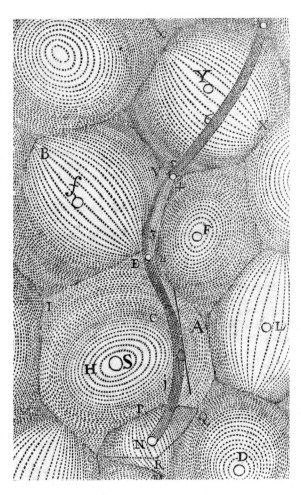

Descartes' model of the universe as a series of abutting vortices allowed both movement and continuous matter.

The original atomists might have argued that it was impossible for matter to be both particulate and continuous, for particles could only move if they had somewhere to move to, and that required empty space. Descartes' model was more like a tank of fish: the water is continuous, the fish are discrete but continuous within themselves (and, to all intents and purposes, are indivisible particles), yet they can move through the water. Descartes' vision of the universe was characterized by vortices, with matter moving in circles (or spheres) so that no empty space was needed (see illustration on opposite page).

For Descartes, there were only two fundamental (or primary) qualities of matter: extension in space and movement. Although he accepted a corpuscular model, Descartes denied that corpuscles were indivisible. He used an argument that most people would now consider specious: Descartes claimed that even if we can't divide particles beyond a certain point, God can, and even if God chooses not to, the fact that He could divide them if He wanted to means they are, by definition, divisible.

Shapely atoms

Gassendi needed no such mental gymnastics. He was happy to entertain a void in which atoms could move around. He argued that all material substances must be composed of elemental particles which share some common features essential to matter, and that atoms are best suited for carrying those features. Following an ancient argument, he claimed that atoms must be hard because if they were soft there could be no hard matter; even cramming them right up against one another would still produce soft matter. If they were hard, soft matter could be constructed by leaving space between the atoms to allow some 'give'.

Gassendi distinguished between the properties of matter that were produced by atoms, and the properties that emerged when atoms were grouped together. He

considered the inherent properties of matter to be solidity, size, shape and weight. He assigned the four Greek elements of hot, cold, wet and dry to atoms of different types. Heat atoms were small, round and fast-moving, while atoms of cold were spiky and pyramid-shaped (reflecting the fact that cold can feel sharp). Gassendi considered magnetism, heat and light all to be atomic in make-up.

The spiky shapes which Gassendi assigned to cold are replicated in ice crystals - cold solidified.

In Gassendi's model, atoms came in a relatively small number of sizes and weights, but a large (though finite) number of shapes. This explained all the different types of matter we observe in the world around us. Atoms, Gassendi said, have the ability 'to take hold of each other, to attach themselves to each other, to join together, to bind each other fast.'

Gassendi drew on the development of the barometer in the 1640s (see page 51) and the way in which it demonstrated the existence of a vacuum to support his model of atoms in a void. He explained the existence of vapour in terms of atomism – atoms moving further apart create more of a void between them. Regarding solubility, he explained that silver could be dissolved in *aqua fortis* (nitric acid) and gold in *aqua regia* (a mixture of nitric and hydrochloric acids) because in each case the shape of the atoms of metal fitted exactly into the 'pores' in the liquid.

Evangelista Torricelli experimenting with a barometer in the Alps in 1643.

Parts and pores

The Ancient Greeks first suggested that matter has substance and pores. Aristotle accepted it, since pores needn't necessarily be voids, they just make matter a complex shape. They could explain how substances react together, as in Gassendi's account of solubility, how rigid or soft a substance is, its melting point (atoms of heat can worm into the pores) and many other characteristics.

Isaac Newton suggested that if a body is made of equal proportions of particles and pores (space), and those particles themselves can be broken down into particles and pores, it's easy to approach a situation in which matter is mostly space. This theory matches quite well the modern model of matter, in which molecules are made of atoms in space and atoms are made of subatomic particles in space. The modern model has far more space than matter, whereas Newton suggested an equal amount of space and matter at each step. His intention was to demonstrate how light, magnetism and gravity could pass through apparently solid objects.

Making space

The question of the void would soon be resolved. In 1630, Galileo Galilei suggested the reason a siphon could not raise water above 10m (33 ft) was because a vacuum held the water up. So Gasparo Berti set out to test whether a vacuum could be created. He took a lead tube measuring 11m (36ft)

THE POWER OF NOTHING

The question of continuous or discontinuous matter seemed to be answered in 1654. In that year, German scientist Otto von Guericke staged a series of demonstrations which showed not only the immense power of atmospheric pressure but also that it is possible to have (or create) a void. He had made an air pump that was capable of creating a vacuum in a container. His dramatic demonstrations consisted of removing the air from a pair of metal 'Magdeburg hemispheres' which had been fitted together. Von Guericke then showed it was impossible to pull them apart. Even teams of horses could not separate the two halves because the pressure of the outside air held them tightly together.

in length, sealed one end and filled it with water. He inverted it in a container of water. Some water flowed out of the tube, leaving an apparently empty space at the top.

Evangelista Torricelli repeated the experiment using a heavier liquid. Taking a glass tube about 1m (3ft) long he closed one end, filled it with mercury and inverted it in

The Magdeburg hemispheres were not the first evidence of atmospheric pressure though they did constitute a stunning demonstration of just how much pressure the air around us exerts.

a dish of mercury. The level of the mercury in the tube fell to around 76cm (30in). As atmospheric conditions changed, the level rose and fell. Torricelli had invented the barometer. The science is easily explained now: if the external air pressure drops, the mercury falls in the tube as there is less pressure acting on the reservoir outside the tube. If the external air pressure rises, the increased pressure on the reservoir forces more mercury up inside the tube.

Torricelli's results still met with considerable resistance. Some people argued that the space at the top of the tube wasn't a vacuum, but was probably filled with vapour from the mercury. The French mathematician Blaise Pascal repeated the demonstration in 1646 using wine and then water, and compared the two. The volatility of the alcohol in the wine should have meant that, if evaporation was taking place, more vapour should be present in the wine tube than in the water tube, but there was no difference in the height of the two columns under the same conditions.

Soon after, an English chemist turned his attention to air and the pressure it can exert. Robert Boyle (see box, below) is important in the story of the Periodic Table for his work on gases and their particulate nature, and his thoughts on the elements. He was the first chemist to reject the Aristotelian and Paracelsian versions of the elements and argue for something closer to the modern paradigm, based on the nature of the particles making up matter.

ROBERT BOYLE (1627–91)

Robert Boyle was born in Ireland, the fourteenth child of the Earl of Cork, and received his early education there before being sent to Eton College in England. At the age of 11, he set off on a six-year educational tour of Europe, finally settling in Dorset, England, in 1649. There he began writing, and set up a laboratory where he could pursue his scientific investigations. In 1655 he moved to Oxford, where he met other natural philosophers, including Robert Hooke. He was a founder member of the Royal Society, which began as an 'invisible college' in 1660 and received its royal charter and current name in 1663. He published his thoughts on the 'corpuscular' nature of matter in *The Sceptical Chymist* in 1661 (see page 74), the year after he had produced his even more famous work stating Boyle's law, which laid the foundation of pneumatics.

Boyle was keen to raise the status of chemistry and to dissociate it from the commercial and medical concerns of the 'chymists' of his day. (Chymists represented a sort of halfway house between alchemists and the practitioners of modern empirical chemistry.) As the son of the wealthiest man in Britain, he was at license to pursue his chemical experiments and entertain the prominent scientists of the day.

The 'spring of the air'

Boyle had read about Guericke's air pump and set out to make his own improved design. He used it for several experiments, including one famously depicted in a painting by Joseph Wright of Derby, *An Experiment on a Bird in the Air Pump* (above).

The results of Boyle's experiments with the air pump and with air pressure led to the publication, in 1660, of *New Experiments Physico-Mechanicall, Touching the Spring of the Air and Its Effects*, and to the formulation of his law. Boyle's law states that the pressure exerted by a gas at a

Above: Joseph Wright's An Experiment on a Bird in the Air Pump, *1768, depicts a demonstration in which Boyle removed the air from a glass dome containing a bird, causing the creature to die of cold.*

fixed temperature is inversely proportional to the volume it occupies, or:

pressure × volume = constant

This means that if a gas is compressed, its pressure increases, and if it is allowed to spread out, occupying a greater volume, the pressure decreases. Boyle conceived of air as tiny particles separated in space by miniature springs that could be compressed

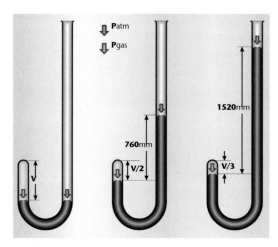

Left: Boyle experimented with the pressure and volume of air by trapping a small amount of air in a J-shaped tube behind a column of mercury. As he added more mercury to the open end on the right, he measured the change in volume of the trapped air on the left.

or expanded. The particles could be squashed into a small space if the springs were compressed; if the springs were relaxed, the particles could expand to fill the space that became available.

Air and nothing

Boyle was adamant that his air pump removed everything from the glass dome. But detractors, such as the political philosopher Thomas Hobbes, claimed that something, perhaps *aether*, remained in the flask. For these refuseniks, the idea of a void was inconceivable and no amount of proof would sway them. For others, though, the vacuum pump refuted the Aristotelian objection to the idea of a void (or vacuum). If the air could be expelled from a container so that the container held nothing at all, matter was clearly not continuous. It was a crucial step on the path to understanding the elements. The possibility of atoms existing and moving in a void, combining and recombining, meant that the model at the centre of modern chemistry could begin to emerge.

Boyle's alchemy

However, it would be wrong to think that Boyle was carrying chemistry towards a glorious future divorced from alchemy. Boyle was not a skeptical alchemist; in fact he was so thoroughly persuaded of the veracity of the demonstrations of alchemy he had witnessed that he helped campaign to overturn the law banning its practice. In this he was successful. In 1689, it became legal in Britain to attempt to transmute base metals into gold. Boyle had no doubt that this was theoretically possible, and actively pursued the quest. He had been influenced (or cheated) by a number of alchemists, and was convinced he had witnessed transmutation himself.

Where Boyle differed from most alchemists, though, was in his rejection of the four Aristotelian elements and the three Paracelsian 'principles' of matter. Although followers of Paracelsus claimed they were able to break down matter into its constituent parts using fire, like other forward-thinking scientists before him Boyle believed that burning actually changed the nature of the material rather than revealed its ingredients. Their methods, he argued, did not uncover something hidden, but produced the substances they claimed to have found.

Particles matter

Boyle replaced the Aristotelian and Paracelsian elements with a single type of matter which he called 'catholick matter'. He considered that all things were made of this substance. As we have seen, several of the Ancient Greek philosophers had proposed that all substances are in essence a single entity: for Thales it was water and for Anaximenes it was air. In Boyle's own era, Flemish chemist Jan Baptist van Helmont (1580–1644) also suspected water to be the basis of all matter (see page 59). But Boyle's model could be made to work logically in a way that the other models could not.

Boyle accounted for the different properties of matter by supposing that

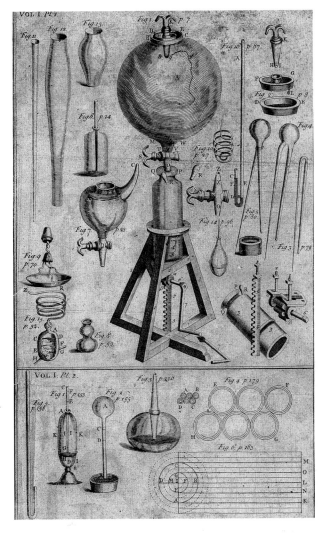

Right: Boyle's air pump: the ability to create a vacuum was crucial to the development of the atomic model of matter.

'There is one catholick or universal matter common to all bodies, by which I mean a substance extended, divisible and impenetrable.

'But because this matter being in its own nature but one, the diversity we see in bodies must necessarily arise from somewhat else than the matter they consist of. And since we see not how there could be any change in matter, if all its (actual or designable) parts were perpetually at rest among themselves, it will follow, that to discriminate the catholick matter into variety of natural bodies, it must have motion in some or all of its designable parts.'

Robert Boyle, *The Origine of Formes and Qualities*, 1666

the tiny corpuscles of 'catholick matter' exist in different sizes and shapes. The size, shape, interactions and movement of these corpuscles produce, he claimed, the diversity of matter in the world around us.

He saw no conflict between his chemical model of matter and the possibility of alchemical transmutation, writing:

'Since bodies, having but one common matter, can be differenced but by accidents, which seem all of them to be the effects and consequents of local motion, I see not why it should be absurd to think that . . . by the intervention of some very small *addition* or *subtraction* of matter, . . . and of an orderly *series* of *alterations*, disposing by degrees the matter to be transmuted, almost of any thing, may at length be made any thing.'

Like Descartes, Boyle accounted for the secondary properties of matter (colour and so on) by the way the shapes of the corpuscles interact with our sense organs. He explained the way that matter can change (in a chemical reaction, for example) and develop different properties by recourse to the same model. If corpuscles aggregate, the compound corpuscles will necessarily have a different size and shape and will interact differently with the sense organs and with other substances:

'as well as each of the *minima naturalis*, as each of the primary clusters . . . having its own determinate bulk and shape, when these come to adhere to one another, it must always happen that the size and often the figure of the corpuscles composed by their juxtapositions and cohesion will be changed. . . . And whether anything of matter be added to a corpuscle or taken from it . . . the size of it must necessarily be altered, and for the most part the figure will be so too, whereby it will acquire a congruity to the pores of some bodies . . . and become incongruous to those of others; and consequently be qualified . . . to operate on divers occasions, much other wise than it was fitted to do before.'

As the clusters or aggregates (what we would call molecules) are made by putting the simplest corpuscles (what we would call atoms) together, their consequent shape and properties belong to the cluster itself. The original components and their qualities can be restored by breaking apart the cluster.

Back to the air

Although Boyle's ideas about the nature of matter will prove important when we turn again to the concept of elements, his work on 'the spring of the air' opened the door for his successors to investigate gases further. With pressure and particles in place, the next step in investigating gases would be the recognition that there is more than just the one 'air'.

A chaos of gases

Suffocation or drowning would have made our need for air obvious even to our very earliest ancestors. Air is invisible, yet we can see the effects of wind moving the trees and we can see smoke and steam carried in the air. Miners experienced (and sometimes died from) the 'noxious' air in coal mines. But people simply couldn't understand the scientific reason for this. Air is difficult to explain if you have no concept of a gas made up of moving particles. Once that idea was in place, the investigation of gases by way of the reactions they are involved in could begin.

Although scientists stumbled across carbon dioxide and hydrogen before the mid-18th century, these gases were not formally 'discovered' until useful observations and measurements could be made of them. In a flurry of activity, between 1766 and 1774, the main atmospheric gases and a couple of others were discovered. These were: nitrogen, oxygen, carbon dioxide, hydrogen and chlorine.

Canaries were used in coal mines in Britain from 1911 until 1986, acting as sentinels to alert miners to dangerous levels of carbon monoxide. The canary would die before the miners were seriously affected, giving them a chance to leave the area.

One air or many airs?

Italian polymath Leonardo da Vinci (1452–1519) may have been the first to notice that something in the air is needed for a candle to burn or an animal to respire. He recorded his observation, alongside his speculation that air contains more than one component, in his unpublished notebooks, so it did not feed into mainstream scientific discourse.

In 1604, Polish alchemist, Michał Sedziwój (1566–1636), or Sendivogius (his Latinized name), noted that air contains a life-giving substance which seems to be the same as the one given off by the decomposition of potassium nitrate (KNO_3). This is true: potassium nitrate breaks down to give potassium nitrite, KNO_2 and oxygen. Sendivogius built a philosophical schema of the universe around this 'food of life' in the air. He worked in a series of royal courts in Poland, Bohemia and Moravia, where rulers with an interest in alchemy were keen to engage him. The practical chemistry of gases was not his main interest.

The next and most significant contributor was Jan Baptist van Helmont. In 1604, he showed that a burning candle uses up something in the air. His experiment demonstrated not only that the composition of the air changes, but that the volume of gas remaining is also reduced. He burned a candle in a glass container inverted over a dish of water and showed that the candle would burn for a while and then go out. Meanwhile the rise in the water level in the container demonstrated that some of the air

The Polish alchemist Sendivogius was more interested in purifying various substances than in his discovery of the gas that would be identified as oxygen.

DOES EVERYTHING COME FROM WATER?

Van Helmont is most famous for his experiment with growing a willow tree for five years to demonstrate that everything can be fabricated from water. He began by weighing the sapling and a quantity of dry soil, which he then put into a pot. He covered the pot to prevent dust falling into it and fed the sapling nothing but distilled water. After five years, he removed the sapling from the pot, weighed it, and dried the remaining soil and weighed that. He concluded that as the weight of the soil changed very little and the weight of the tree changed a great deal, all the matter making up the tree must have come from the water he had provided. The fault in his reasoning, clear to us but not to him, was that he took no account of the gases available to the tree. In order to respire and grow, the tree had used oxygen and carbon dioxide from the air as well as the water provided.

Mayow's experiments with combustion and respiration demonstrated that a mouse and a candle both need and use up a component of the air. As usual, the outcome for the mouse was unfortunate.

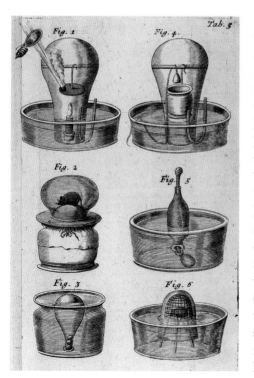

had been used up and the water had risen to take its place.

Van Helmont's experiment was repeated by others more than 150 years before the composition of the air began to be discovered (see page 64). In 1674, English physician John Mayow (1641–79) devised a quantitative demonstration using two glass flasks inverted over water. He placed a mouse in one flask and a lighted candle in the other and measured the rise in water level as the mouse and the candle expired. In each case he found that about $1/14$ of the volume of air in the flask had been used up. There were two important conclusions to be drawn: one was that the mouse and the lighted candle needed the same component in air; the other was that air has more than one component, and the one in question represents about $1/14$ of its volume. He was wrong in the second conclusion; oxygen (the gas at issue) represents about a fifth of the composition of normal air. The mouse and candle expired when the oxygen level fell too low rather than when all the oxygen had been used up.

CHOKE-DAMP, BLACK DAMP, WHITE DAMP AND STINK DAMP

The notion that some air is lethally unbreathable was familiar to miners. The terms 'choke-damp' or 'black damp' were used for the unbreathable, oxygen-poor air encountered in coal mines where exposed coal had reacted with the oxygen in the air, producing an atmosphere of nitrogen, carbon dioxide and water vapour. White damp is now known to be carbon monoxide, and 'stink damp' is hydrogen sulphide (which smells like rotten eggs). 'Damp' is from the German *dampf*, meaning 'vapour'.

Joseph Black lecturing on chemistry at Glasgow University, Scotland.

Investigating 'chaos'

Van Helmont was the first person to coin the word 'gas'. Before this, people spoke of different types of 'air' but had no concept of distinctly different gases. Similarly there was no notion of gases that do not form part of the atmospheric air. Van Helmont derived 'gas' from the Greek 'chaos'. (In this sense, chaos referred to the primordial void from which the Ancient Greeks believed matter to have been created, rather than the modern sense of complete disorder.)

Van Helmont was also the first to describe and name a gas other than air. He found that if he burned charcoal, the mass of ash left behind was much less than the mass of the charcoal he had started with: burning 62 lb (28 kg) of charcoal resulted in only 1 lb (0.5 kg) of ash. Van Helmont

suggested that the rest of the charcoal had been turned into an invisible substance, a gas or *spiritus sylvestris* ('wild spirit').

It may seem that van Helmont should get the credit for discovering carbon dioxide (which is what his *spiritus sylvestris* was), but this generally goes to Scottish physician Joseph Black who experimented with carbon dioxide (CO_2) in the 1750s. He found that this gas was produced by respiration and could not support life – though these were hardly new discoveries. He called it 'fixed air'.

Where Black differed crucially from van Helmont was in his demonstration that 'fixed air' could be generated by chemical means. He could produce it by heating calcium carbonate, and it could be involved in chemical reactions. Prior to its release from its chalky prison, it was an integral part of a solid.

This astonishing new insight – that there were multiple gases, and that they could partake in normal chemical reactions and combine with other substances to make solids – prompted a period of intense investigation that led to the rapid discovery of four more gases, including three that turned out to be elements.

The different airs

While the work of chemists had shown that there are different constituents of air, separate gases were not captured and identified until the middle of the 18th century. Strangely, the first gas to be identified was not one that forms part of Earth's atmosphere. It is, however, liberated readily in chemical reactions, so is easily collected in experiments.

Schweppes has continued to make money from carbon dioxide for more than two centuries.

FROM FIXED AIR TO FIZZY WATER

In 1772, English chemist Joseph Priestley published a paper describing how to make 'fixed air' by dripping sulphuric acid (oil of vitriol) on to chalk. He also explained in 1770 how to make fizzy water by dissolving the gas. Using the method he described, a German-born watchmaker from Switzerland called Johann Schweppe began to manufacture carbonated water. His business in London failed in 1795, but Erasmus Darwin (grandfather of Charles) popularized the drink. The beverage soon became a success after King William IV was said to enjoy drinking it. The Schweppe company still makes fizzy drinks today.

Burning bright

Around 1700, French chemist Nicolas Lémery dissolved iron in sulphuric acid and put a lighted candle to the gas released. When he discovered that it produced a violent fire and noise, he thought he had discovered the source of thunder and lightning. In fact he had found but not recognized hydrogen.

In 1766, English chemist Henry Cavendish (1731–1810) isolated and identified hydrogen, which he called 'inflammable' air. In a closed flask, he dissolved zinc in concentrated hydrochloric acid and trapped the gas produced. When he lit it with a splint, it burned explosively. Cavendish believed that the hydrogen came from the metal in his reaction, rather than from the acid. Hydrogen had already been produced by Paracelsus, Boyle and another English chemist, Joseph Priestley, but it was Cavendish who identified it as a separate gas. (Lavoisier named it hydrogen.) Cavendish discovered that burning hydrogen in air produces water.

Hydrogen occupies a special place in the Periodic Table as the first element. It is the most abundant element in the universe, even though it doesn't exist for long in its native form on Earth (see page 133).

METALLIC HYDROGEN

Under immense pressure, hydrogen is thought to condense to a thin, reflective liquid – metallic hydrogen. The single electron of a hydrogen atom is freed to flow through the substance, making the hydrogen a conductor of electricity.

On Earth, metallic hydrogen has possibly been observed under pressure around 500 million times that of Earth's atmospheric pressure, though the validity of the claim (made in 2017) is contested.

NASA experts believe a sea of metallic hydrogen is hidden beneath Jupiter's gas clouds, and that it could be over 40,000 km (25,000 miles) deep.

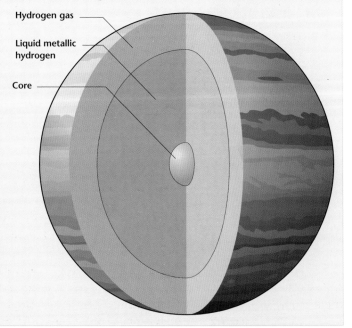

Hydrogen gas

Liquid metallic hydrogen

Core

The English chemist Henry Cavendish first identified hydrogen as a separate gas.

1734). His idea was that phlogiston is present in everything flammable, and the process of burning frees the phlogiston into the air.

The argument went that when something was burned in air or when an animal respired, phlogiston was released into the air. When the air was completely saturated with phlogiston and could take no more, burning (or respiring) would stop. This is the opposite of the reality, of course, in that the burning and respiring stop when the oxygen has been largely exhausted – something has been removed from the air rather than added to it.

It was in the light of this widely accepted model that the discoveries of hydrogen and oxygen were made and framed. When Cavendish liberated hydrogen, he at first thought it was the principle of flammability, then phlogiston, and then a mixture of phlogiston and water.

A burning issue

Combustion was a key process in isolating hydrogen and, later, oxygen. The latter was most easily identified as the product of a reaction by burning something in it – or by using a hapless mouse to try breathing the gas produced in a reaction. But the issue of combustion was contentious. Many chemists had noted that most matter loses mass when it is burned – but, oddly, metals often gain mass when heated in air. The prevailing explanation for the loss of mass was that most matter contains a combustible element, named 'phlogiston' by German chemist Georg Stahl (1660–

The food of life

Oxygen, the gas used up by candles and mice, which Sendivogius had called the 'food of life', was discovered around 1772 by Swedish apothecary Carl Wilhelm Scheele (1742–86). He called it 'fire air' after the brilliant sparks created when he exposed red-hot manganese oxide to hot charcoal dust in the presence of this 'air'. He discovered he could make the same 'fire air' by heating potassium nitrate, mercuric oxide, or many other substances.

Although Scheele carried out many experiments, creating 'fire air' in various reactions and documenting them carefully, he didn't publish his findings until 1777. By that time Joseph Priestley had also discovered the gas. Priestley found in 1774 that if he heated mercuric oxide in a sealed container, it produced a gas that was 'five or six times as good as common air' at sustaining a candle or live mouse. He considered it to be 'dephlogisticated air'– air from which all the phlogiston had been removed. His special air could, logically, support burning or breathing for longer than normal air before becoming saturated. Even though he was wrong in his interpretation, the implications were important: 'Air is not an elementary substance, but a composition,' he wrote. In his view, air must contain at least two portions: – its breathable part and some phlogiston.

Because he published first, Priestley generally gets the credit for isolating oxygen and recognizing it as being important in combustion and respiration (though his account of its role was entirely wrong).

Noxious nitrogen

Throughout history, people have been breathing in air that contains 78 per cent nitrogen, yet this gas wasn't discovered – or its existence even suspected – until the second half of the 18th century. Cavendish wrote: 'In all probability there are many kinds of air which possess this property [that is, 'air which suffocates animals']. I am sure there are two, namely, fixed air & common air in which candles have burnt.' Of course, 'common air in which candles have burnt' does contain a certain amount of

> *'My reader will not wonder that, after having ascertained the superior goodness of dephlogisticated air by mice living in it . . . I should have the curiosity to taste it myself. I have gratified that curiosity by breathing it, drawing it through a glass syphon, and by this means I reduced a large jar full of it to the standard of common air. The feeling of it in my lungs was not sensibly different from that of common air, but I fancied that my breast felt peculiarly light and easy for some time afterwards. Who can tell but that in time this pure air may become a fashionable article in luxury. Hitherto only two mice and myself have had the privilege of breathing it.'*
> Joseph Priestley, 1775

'fixed air', but more importantly it contains little oxygen. The larger component of this exhausted 'common air' would soon be explored.

Joseph Black, who named carbon dioxide 'fixed air', suggested to a young research student that he might like to study an effect he had noticed: when he burned a carbon-rich substance in a sealed container, and then removed the resulting 'fixed air' produced by absorbing it with caustic potash (potassium hydroxide, KOH), some 'air' remained. In fact, Priestley had already noticed as much and observed that the residual 'air' had 'acquired new properties'. It was slightly lighter than common air. Nevertheless he had not pursued the matter.

The field was free for Black to suggest to Daniel Rutherford that he investigate the properties of this air as part of the work for his doctoral thesis.

In 1772, Rutherford found that a mouse died if put in a flask containing this air, so he called it 'noxious' air. We call it nitrogen. He didn't recognize it as a constituent of the atmosphere, but considered it to be completely phlogisticated air (even more phlogisticated than fixed air). He had suffocated a mouse in a contained volume of air, then burnt a candle in the remaining air, followed by a piece of phosphorus, removed the resultant fixed (or 'mephitic') air, and then tested the remaining gas: 'Healthy and pure air, by being so respired, not only becomes partly mephitic, but also suffers another change in its nature. For after all the mephitic air

Liquid nitrogen boils at -196° C (-321° F) and is widely used as a refrigerant. Nitrogen gas is used in the mining industry to put out fires, as it quickly combines with oxygen to form nitrous oxide, removing the oxygen from the air and starving the fire.

is separated and removed from it by means of a caustic lixivium, that which remains does not thence become more healthful; for although it makes no precipitate of lime from water, yet it extinguishes fire and life no less than before.'

The fixation on phlogiston not only complicated the way that chemists thought about gases, it threatened to derail the entire endeavour. Fortunately, a more cynical and ambitious chemist was waiting in the wings. Phlogiston would become a casualty of

NITROGEN – ESSENTIAL POISON

Although the early chemists noted that nitrogen could not sustain life as an 'air', nitrogen and the nitrogen cycle are essential to life on Earth.

Nitrogen is 'fixed' from the air by bacteria in the soil. It is absorbed by plants which are eaten by animals or which die and decay, releasing nitrogen back into the soil and air through the action of denitrifying bacteria. The nitrogen-rich plants are eaten by animals and the nitrogen moves through the food chain, with some being lost in excrement on the way. Eventually the animals die, releasing nitrogen into the soil and atmosphere. In the living bodies of plants and animals, nitrogen is a vital component of proteins and DNA, which carries the genetic information describing all living things.

To sustain the current level of human population it is essential to use nitrogenous fertilizers to promote plant growth. These were first sourced from night soil, bone meal and guano (harvested bird droppings), but since the early 20th century nitrogen from the atmosphere has been fixed into ammonia by means of the Haber process, developed by two German chemists, Fritz Haber and Carl Bosch.

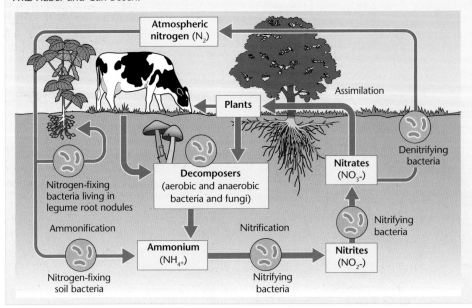

a new approach to the science promoted by the man often considered the father of chemistry, Antoine Lavoisier.

Visible air

The last gas to be found for many years was discovered by Scheele in 1774. Chlorine was the first clearly visible gas to be isolated. Scheele produced it by creating a reaction between a mineral called pyrolusite (manganese dioxide) and hydrochloric acid. The result was a choking greenish-yellow gas which dissolved in water to produce hydrochloric acid. Scheele was convinced that the gas contained oxygen. It was not identified as an element until 1810, when English scientist Humphry Davy did so and named it chlorine from the Greek *chloros*, meaning greenish-yellow. Not all chemists immediately accepted that it was an element, however.

Life and fire unlocked

In the same year that Scheele discovered chlorine, Joseph Priestley visited Lavoisier in Paris. Lavoisier was already suspicious of the phlogiston model of combustion, and what Priestley told him confirmed his opinions.

Chlorine is produced industrially from sea water, using electrolysis to free it from salt (sodium chloride) dissolved in the water.

Lavoisier had learnt of phlogiston theory when he attended lectures on science while supposedly studying for a law degree. He knew that some substances gain rather than lose mass when burned. These substances include the metals that combine with oxygen to form calces (oxides). Boyle had explained this by suggesting that corpuscles of heat worm their way through the glass of a container and embed themselves in the metal, so increasing its mass. Some scientists even suggested that phlogiston might have negative mass. But this looked to Lavoisier like a fudge. He had discovered that burning sulphur and phosphorus gained mass by combining with air; he also knew that if he heated a metal calx he could trap a large amount of air released by it. He believed combustion could probably be accounted for by science, without invoking such an unlikely substance as phlogiston.

67

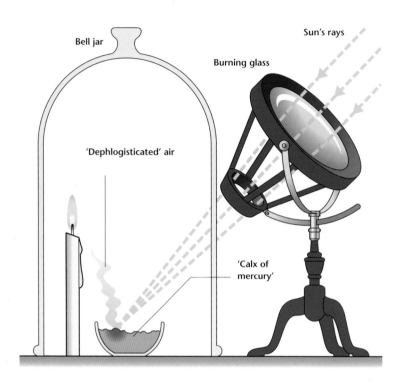

Priestley's bell jar experiment.

Priestley told Lavoisier that he had heated mercury calx and captured a gas in which a candle burned longer than it did in ordinary air. Priestley's interpretation was that the gas he had collected was free from phlogiston; he called it dephlogisticated air. Lavoisier repeated Priestley's experiment and came to the conclusion that air contains at least two components. He termed one of these 'respirable air', which was necessary to support life. It was also involved in combustion, which Lavoisier described as a reaction between carboniferous substances or metals and respirable air. Taking this component out of the air to form a compound made the air unsuitable for breathing or further combustion. He publicized his new theory of combustion in 1777.

In 1779, Lavoisier announced to the Royal Academy of Sciences in Paris that most acids contain breathable air. He named it *oxygène*, from the Greek for 'acid generator'. This completely revolutionized chemistry; the debunking of phlogiston freed the science from a major obstacle to its progress. The start of modern chemistry can plausibly be dated from this moment.

> 'It took them only an instant to cut off this [Lavoisier's] head, and one hundred years might not suffice to reproduce its like.'
> Joseph-Louis Lagrange, Italian mathematician and astronomer, on the execution of Antoine Lavoisier

ANTOINE-LAURENT LAVOISIER (1743–94)

The son of a wealthy Parisian lawyer, Lavoisier followed his father and trained in law but chemistry was his true passion. He married Marie-Anne Pierrette Paulze in 1771 when she was only 13 and he was 28. She would become a skilled and dedicated professional ally and companion.

From 1775, Lavoisier worked as commissioner of the Royal Gunpowder and Saltpeter Administration at the Paris Arsenal, where he improved the methods used to manufacture gunpowder and attracted chemists from around Europe to his well-equipped laboratory. He worked extensively on gases, debunked the phlogiston theory (see opposite page), stated and applied the principle of the conservation of matter, and introduced the modern system of naming chemicals. He investigated the adulteration of black-market tobacco and suggested adding a little water to state-approved tobacco to improve the taste. Most importantly (for our purposes) he set out the defining feature of a chemical element and compiled the first modern list of elements. His careful measurement and documentation and rigorous attention to detail made him an exemplary scientist. He published his seminal work *Traité élémentaire de chimie* (*Elementary Treatise on Chemistry*) in 1789, setting out to revolutionize chemistry and drag it into the modern age.

But 1789 was also the year that marked the start of the French Revolution. As a wealthy intellectual, Lavoisier was in extreme danger. He was arrested in 1793 and accused of abusing state funds and adulterating tobacco. He was executed by guillotine in 1794. His seized goods were returned to his widow the following year, with a note acknowledging that he'd been falsely convicted.

Marie-Anne Lavoisier taught herself English so that she could translate scientific papers for her husband, learned science so that she could understand his work, and became adept at illustration and engraving so that she could illustrate his publications.

Two elements down: air, then water

Just as Priestley's work contributed to Lavoisier's recognition of the nature of air and the non-nature of phlogiston, Cavendish's work contributed to Lavoisier's understanding of water.

Cavendish had found that if he burned hydrogen in air, water was a product. He concluded that water was present in both types of air at the start and was released in a reaction that he formulated to follow the phlogiston model.

Lavoisier explained the process in terms of his own model, in which combustion occurs in combination with oxygen. In 1783, he burned hydrogen in oxygen, gaining very pure water, and came to the entirely logical – even inevitable – conclusion that burning in this case involved hydrogen combining with oxygen to produce water. Water was not, then, an element at all, but a compound of hydrogen and oxygen. To confirm his conclusion beyond doubt, he decomposed water into hydrogen and oxygen by passing water vapour over red-hot iron filings and collecting the gases produced over mercury. They were, as he expected, hydrogen and oxygen. This result gave him the confidence to launch a full-scale assault on phlogiston. He denounced it as 'imaginary' and 'a veritable Proteus that changes its form every instant'. This was not a universally popular move, and many scientists – including Priestley – clung to the old model until their dying day.

The final stage of Lavoisier's investigation into the composition of water aimed to discover the proportions of hydrogen and oxygen in water. He carried out an ingenious experiment. He added a little water and some iron filings to a small beaker of mercury. As both water and iron are lighter than mercury, they floated. He then inverted the beaker in a larger bowl of mercury, so the water and iron were trapped near the top. He left the apparatus for a few days, during which time the iron reacted with the water to produce iron oxide and released hydrogen, which collected above the mercury. From his previous experiment, Lavoisier knew the volume of gases to expect from decomposing a given volume of water. By measuring the volume of hydrogen produced in this experiment, he was able to calculate that water consists of two thirds hydrogen and one third oxygen.

The end of the four

By 1783, Lavoisier had demonstrated conclusively that neither water nor air is an element. Van Helmont's experiment with the willow sapling had suggested that earth is not an element either, or at least it is not needed in the construction of a willow tree.

In a further experiment, Lavoisier disproved the idea that water could be transmuted to earth by evaporation. This belief rested on the observation that if water is set boiling in a glass flask for long enough, there is an 'earthy' sediment when all the water has evaporated. By careful measurement, Lavoisier showed that the 'sediment' was actually the product of disintegration of the glass vessel through the process of extended reflux heating.

A new chemistry

For Lavoisier, the debunking of phlogiston was the first step in ridding chemistry of its accumulated errors and elevating it to a professional standing and rigorous scientific consistency. He had abandoned law for chemistry in 1772 and had begun to revolutionize the discipline in 1783. Eleven

By burning hydrogen in oxygen to produce water, Lavoisier proved conclusively that water is not an element, but a compound.

years later he would be dead, executed by guillotine during the French Revolution. In these few years, he had much important work to do.

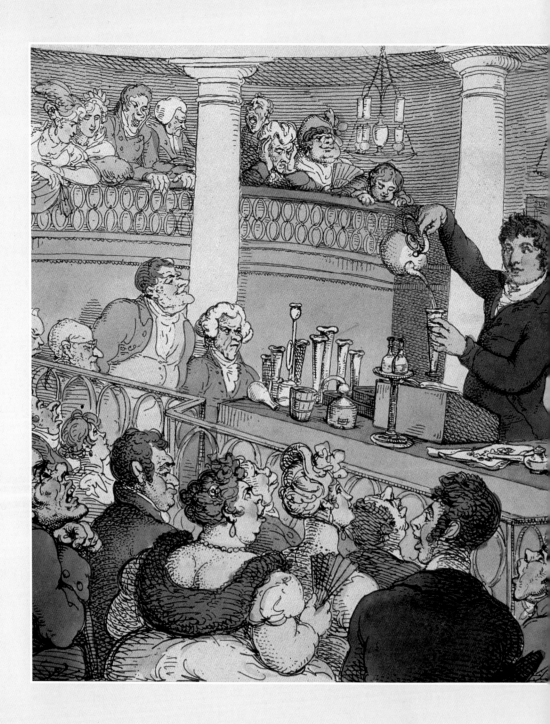

CHAPTER 4

THE NEW ELEMENTS

'I now mean by Elements, as those Chymists that speak plainest do by their Principles, certain Primitive and Simple, or perfectly unmingled bodies; which not being made of any other bodies, or of one another, are the Ingredients of all those call'd perfectly mixt Bodies are immediately compounded, and into which they are ultimately resolved.'

Robert Boyle,
The Sceptical Chymist, 1661

In the 18th century, the recognition that both air and water can be broken down into at least two gaseous ingredients was the final nail in the coffin of the four Greek elements. A new model emerged which was the basis of the modern conception of the elements and brought us a little closer to the Periodic Table.

Thomas Rowlandson's satirical cartoon Chemical Lectures, *1808, reflects the popular interest in chemistry in the early 19th century.*

Boyle and the elements

Robert Boyle rejected both the Aristotelian elements of earth, water, air and fire and the Paracelsian principles of salt, mercury and sulphur. He claimed, quite reasonably, that it is impossible to make any substances using the Greek 'elements' and similarly impossible to reduce any substance to them.

Boyle speaks of 'perfectly unmingled bodies' which are the 'Ingredients of all those call'd perfectly mixt Bodies'. This sounds remarkably close to the modern

> 'The importance of the end in view prompted me to undertake all this work, which seemed to me to be destined to bring about a revolution in . . . chemistry. An immense series of experiments remains to be made.'
>
> Antoine Lavoisier, notebooks, 1773

definition of elements ('unmingled bodies') and compounds ('mixt Bodies'). The contemporary notion of corpuscularism held that matter was made up from aggregates of corpuscles and it was from the size, shape and movement of the corpuscles and aggregates that the features of matter derived. The 'mixt Bodies', then, were aggregates of corpuscles and made up all the matter we observe around us. This is not the same as having discrete atoms of different substances. Boyle proposed a single 'catholick matter' of which everything is made. This suggested that it is simply the arrangement, size, shape and movements of particles of fundamental matter which bestow the properties of different substances. Although it's difficult to know exactly what Boyle intended to be inferred from this, it is significant that he didn't propose any substances as elements.

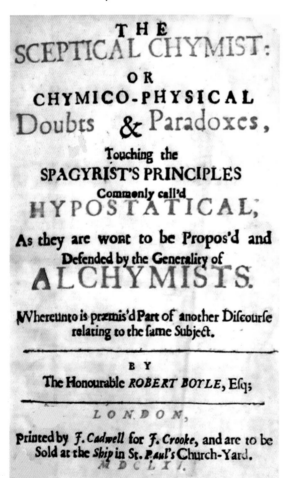

The title page of the first edition of Boyle's Sceptical Chymist, *published in 1661. It was his seminal work, and the text that marks the beginning of a modern approach to chemistry.*

The Chemical Revolution and Lavoisier's elements

In his *Traité* of 1789, Lavoisier defined an element, or 'principle', as a chemical substance that cannot be broken down by any (then-known) method of decomposition. Aware that methods of analysis might one day render some of his elements non-elemental, he said that since 'we have not hitherto discovered the means of separating them' they appeared to be simple substances and should be treated as such 'until experiment and observation' prove otherwise.

Lavoisier listed 33 substances that could not be further broken down. Of that number, 23 are elements but two – 'caloric' (heat) and light – are not even substances. 'Caloric', a hypothetical weightless fluid, might sound suspiciously like phlogiston in concept, but it was very different: Lavoisier considered it to be a massless substance which causes other substances to expand when heated. The idea was that as you heat something, you add caloric to it. Caloric has no mass of its own but takes up space and forces the heated matter apart, increasing its volume.

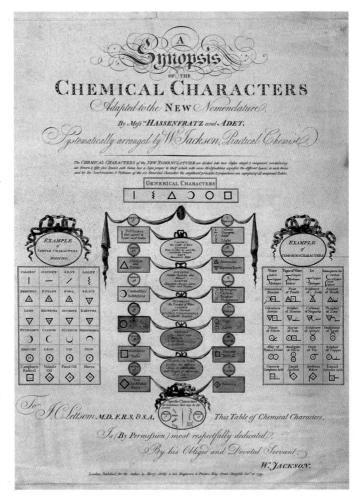

A table of symbols for the elements, produced in 1799 by W. Jackson 'Practical Chemist', includes a symbol for the hypothetical substance 'caloric'.

(Before dismissing the idea, it's worth remembering that the modern concept of dark energy offers a similar explanation for the expanding universe.)

Lavoisier divided his elements into 'elastic fluids', metals, non-metals and 'earths'. The elastic fluids were the known

Boron, discovered in 1808, was predicted by Lavoisier, who identified it as the 'root' of boric acid – the part that isn't hydrogen.

gases – hydrogen, oxygen and nitrogen – plus caloric and light. There were 17 'metals', which he described as 'metallic, oxidizable and capable of neutralizing an acid to form a salt'. Those he listed are all elements: silver, bismuth, cobalt, copper, tin, iron, manganese, mercury, molybdenum, nickel, gold, platinum, lead, tungsten, zinc, arsenic and antimony (though the final two are now considered metalloids rather than true metals). The non-metals were 'oxidizable and acidifiable non-metallic elements'. He counted three true elements, phosphorus,

sulphur and carbon, and substances he called the 'roots' of boric, hydrochloric and hydrofluoric acid. These were later identified as boron, chlorine and fluorine, but had not been isolated at that point. It is impressive that Lavoisier recognized the presence of an unknown element which lay just beyond the abilities of contemporary chemists to isolate. Finally, the substances he listed as earths ('salt-forming earthy solids') are not elements at all. He included the oxides of calcium, magnesium, barium, aluminium and silicon in this category.

An advance on four . . .

Lavoisier's list of 33 elements is a considerable advance on the four that the ancients proposed and the three favoured by the alchemists. It put rather a different complexion on things: if there were to be so many elements (and perhaps more) the task of understanding how matter is put together would become much more complex. But also, perhaps surprisingly, it was more likely to be accomplished. Instead of nebulously perfect or spiritual versions of matter, such as archetypal air and earth or even Boyle's 'catholick matter' embued with qualities, it was now possible that all substances were made by mixing together or in some way fusing simpler but entirely normal forms of matter.

From 33, the number of elements increased rapidly – alarmingly rapidly for many chemists, who struggled to grasp the abundance of elements confronting them.

Emerging elements

Lavoisier's list of elements included several not known to the ancients. Only three elements had been identified between the time of the Romans and that of Robert Boyle. Of course, as no one knew that the substances being discovered were elements, they each gained attention according to their usefulness or curiosity value alone.

It's generally accepted that no elements were discovered between the time of the Romans and AD800, when Arab alchemist Jabir in Hayyan was credited with isolating arsenic

and antimony. After a considerable gap, alchemists were also responsible for the next two elements to be discovered.

Bismuth has been known since ancient times but it was not until the 16th century that it was recognized as a separate metal similar to but distinct from tin and lead (see page 78–9). Around 1669, phosphorus was discovered. Its name, taken from Greek mythology, means 'light bearer' and it is considered the first element of the modern age (see page 79).

Cups made from antimony became popular in the 1600s as a means of purging the body of illness through the sweating and vomiting this poison caused. Wine was allowed to stand in the cup for up to 24 hours, by which time a small amount of the metal had dissolved in it.

The first of few

Bismuth was discovered around 1400 by an unknown German alchemist. It looks spectacular, with crystals with right-angled vertices and a multicoloured iridescent sheen formed by a thin film of oxide over the surface. Bismuth is diamagnetic, which means it produces a field in the opposite direction to any magnetic field applied to it. It's possible to make a magnetic metal float in mid-air by trapping it between conflicting magnetic forces using bismuth.

Bismuth was easily and often confused with lead, being heavy, similarly coloured and with a low melting point. Paracelsus mentions it, though he called it a type of antimony. Decorative bismuth objects exist from the 16th century, but the metal really came into its own with the development of

A powerful rare-earth magnet is trapped between two plates of bismuth with a strong magnet above the top one. The bismuth creates a magnetic field in the opposite direction, causing the rare-earth magnet to hover between the bismuth plates.

printing: it was added to lead to make an alloy for casting type.

While Europeans were beginning to use bismuth in printing, the Incas of South America were using it in bronze. Instead of extracting it from an ore, they apparently mined it in its native form. An Inca knife dating from around 1500 uses tin-copper bronze for the blade and bismuth-copper for the llama-head handle to give a contrasting colour.

In the early 16th century, German mineralogist Georgius Agricola (Georg Bauer, 1494–1555) attempted to categorize bismuth by saying that it was not lead or tin, but 'a third one'. At the time, there were believed to be three types of lead which we now recognize as lead, tin and bismuth. Mineralogists believed that metals slowly transformed from one type to another in the ground (a belief that made alchemical transformation seem all the more plausible). On the transmutative scale, tin was closer to becoming silver than lead was, and bismuth was closest of all. Miners finding bismuth lamented that they had discovered it too early in the transmutation process. Their theory seemed to be supported by the fact that silver was often found just below the bismuth layer. Miners sometimes referred to the bismuth seam as the 'roof of silver'.

Bismuth had no known uses, so miners tended to discard

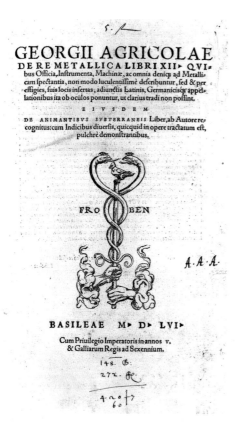

GEORGII AGRICOLAE
DE RE METALLICA LIBRI XII▶ QVI▪
bus Officia, Instrumenta, Machinæ, ac omnia deniᶜᵍ ad Metalli-
cam spectantia, non modo luculentiſſimè deſcribuntur, ſed & per
effigies, ſuis locis inſertas, adiunctis Latinis, Germanicisᶜᵍ appel-
lationibus ita ob oculos ponuntur, ut clarius tradi non poſſint.

EIVSDEM
DE ANIMANTIBVS SVBTERRANEIS Liber, ab Autore re-
cognitus: cum Indicibus diuerſis, quicquid in opere tractatum eſt,
pulchrè demonſtrantibus.

BASILEAE M▶ D▶ LVI▶
Cum Priuilegio Imperatoris in annos v.
& Galliarum Regis ad Sexennium.

Above: The title page of Agricola's treatise on metals and mining, De re metallica, *1556.*

bismuth-containing minerals were not ores of antimony, lead or tin. Soon after this discovery, around 1780, 'bismuth mixture' began to be used to treat gastric disorders, particularly peptic ulcers.

Pee for phosphorus

Phosphorus was the first element to be discovered by an experimental chemist. Hennig Brand (1630–1710) built a laboratory beneath his house in Hamburg where he carried out experiments in his quest to make the philosophers' stone. One process involved collecting large quantities of urine. An account of his method begins, 'Take a good large Quantity of New-made Urine of Beer-drinkers' and continues with an instruction to evaporate it down to a sticky dark mess. (When he says 'a good large Quantity', he really means it: his process required five large barrels of urine, about 5,500 litres (1,453 US gallons) to yield around 60 g (2.11 oz) of phosphorus.) He stored his evaporated urine for a while, and

it as worthless. This meant there was no compelling reason to investigate it and (as a supposed variant of lead or tin) it was ignored until 1753. Then French chemist Claude François Geoffroy demonstrated that it was a distinct metal, and that

Right: Bismuth crystals are strikingly beautiful, with vivid iridescent colours glinting on their tiered rectangular faces.

mixed it with something, probably charcoal, before heating it again and driving off the 'fractions' (that is, separating out substances according to their different boiling points). When this had been done, and when the liquid that was left in his flask was at the requisite heat, yellowish-white fumes passed from it and condensed. Brand had not made the philosophers' stone, he had produced something quite astonishing (and smelly). Phosphorus, which glows in the dark, would have been unlike anything he had seen before. He called it 'cold fire'.

Phosphorus takes several differently coloured forms, or allotropes: red, purple, black and white. It was white phosphorus, now known as P_4, that Brand had made. The mysterious and awe-inspiring glow that Brand observed was finally explained in 1974. As P_4 reacts with atmospheric oxygen, transient products – a hydroxide and an oxide – are formed on its surface. As these quickly decay, light is emitted. In fact, the oxidation process can easily get out of hand, and phosphorus can spontaneously ignite in air.

Brand guarded his secret carefully. Two other alchemists, Johann Daniel Kraft and Johann Kunckel, tried to inveigle or buy the recipe from him. He sold some phosphorus to them and eventually revealed that urine was its source. Kraft demonstrated phosphorus around the courts of Europe, allowing people to believe he had discovered it himself. He gave a small chunk to French chemist Nicolas Lemery, who left some of it on a table with the unfortunate result that it was accidentally bundled up with bed linen. A guest later woke up to find the bedclothes on fire!

PHOSPHORUS IS NOT PHOSPHORESCENT
Despite its name, phosphorus is not phosphorescent but luminescent. Phosphorescent objects absorb light and then emit it again. An energy change in a chemical substance results in the luminescence (or bioluminescence) witnessed in many living creatures, including types of bacteria, fungi, fish, jellyfish and flies.

Above: Bioluminescent jellyfish glow green, a product of a chemical reaction between luciferin and oxygen.

At the same time, Johann Kunckel persevered with the manufacture of phosphorus. By 1676 he was making it fairly easily. He reasoned that if it came from the body it must somehow have got into the body, so he began experimenting with food and other organic materials. At one point he claimed that he could make phosphorus from anything in God's creation, though later he gave up on it, saying: 'I am not making it any more, for much harm can come of it.'

We know of at least one incident which led Kunckel to this conclusion. He had been standing close to a fire with some phosphorus in his pocket, when it ignited spontaneously. His clothes caught fire and his hands were so badly burned that all the skin came off. Although Kunckel rubbed his burning hands in the dirt, he could not quench the fire. He was ill for 15 days. He seems to have been the first victim of something that would become a scourge of the 20th century, when phosphorus became known by some as the 'devil's element' because of the horrific burns it could cause.

In 1678, mathematician and scientist Gottfried Leibniz (famous for discovering calculus independently of Isaac Newton) persuaded Brand to take employment as official alchemist to the Duke of Hanover. Brand showed Leibniz how to make phosphorus. The element continued to be prepared following the same urine-based method until the 1770s, when Scheele found a way of obtaining it from bone.

Joseph Wright of Derby's painting, The Alchemist Discovering Phosphorus, *is thought to refer to Hennig Brand's extraordinary experiment.*

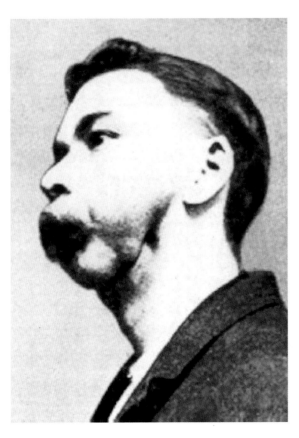

Boyle and the 'spirit of night light'

Robert Boyle witnessed Johann Daniel Kraft's demonstration of the glowing and combustive powers of phosphorus at the Royal Society in London, and was greatly inspired. After considerable experiment and a clue from Kraft that it was made from 'somewhat that belonged to the body of man', Boyle and his

Left: The match industry would claim many unfortunate victims of bone disorders including 'phossy jaw', which eroded the face and jaw. The use of highly toxic white phosphorus in matches was finally banned in various European countries starting in 1872. India and Japan banned it in 1919 and China in 1925. It was never banned in the USA, but punitive taxation was used to make its manufacture unprofitable.

PHOSPHORUS: VITAL AND FATAL

The presence of phosphorus in urine and bone hints at its essential role in living things. Phosphorus is not only necessary as a trace element in bones and so on, it is intricately bound up in every cell of every living organism. DNA, the complex compound that carries genetic information in the form of chromosomes, has a 'backbone' (the two strands of the double-helix) made of phosphate and sugar groups.

Phosphorus was to play a vital role in unlocking the genetic code. In a famous experiment in 1952, geneticists used radioactively tagged phosphorus and sulphur to work out whether genetic information is carried in DNA or in proteins. They exploited the method by which a virus reproduces: injecting its genetic material into a living cell and hijacking the cell's biochemistry to make copies of itself. The result was that tagged phosphorus passed into the cell, but sulphur did not. As phosphorus is present in DNA but not in protein, the role of DNA in heredity was confirmed.

assistant finally produced some phosphorus in 1680. Boyle thought it could have many uses – one of which, he suggested, was to mark the edge of a clock dial so that if he woke in the night he'd be able to see the time. Like phosphorus, clocks that glow in the dark would become the occasion of some terrible injuries and deaths (see page 164). Boyle was the first person to impregnate slivers of wood with phosphorus and use them as matches.

Between Boyle and Lavoisier

Neither Boyle nor Lavoisier discovered any new elements, but between 1661 when Boyle published *The Sceptical Chymist* and Lavoisier's *Traité* in 1789, many completely new elements were found and two familiar metals were properly identified, having previously been thought to be versions or alloys of other metals. French chemist Claude François Geoffroy showed bismuth to be neither tin nor lead in 1753, and platinum was conclusively distinguished from silver around 1749. The gaseous elements hydrogen, oxygen and nitrogen were isolated and identified between 1766 and 1772. New metals that came to light during Lavoisier's lifetime include magnesium, manganese, molybdenum, nickel and tungsten.

White gold

Platinum is often found with gold deposits in Ecuador and Colombia and was used by the indigenous peoples of South America. It's sometimes found as inclusions in gold artefacts, but was also used specifically on its own. As platinum has a very high melting point (1,770°C), it would have

ANOTHER GLOWING STONE

Phosphorus is not the only material that glows naturally, nor was it the first to be found. In 1603, shoemaker and would-be alchemist Vincenzo Cascariolo was digging through volcanic rocks near Bologna, Italy, when he uncovered a white stone. He took it home to experiment with. After heating it, he found that if exposed to light it would glow in the dark for many hours. What Cascariolo had found was a form of barite, or baryte (barium sulphate), the first phosphorescent material. Barite is a compound of barium, sulphur and oxygen, but with impurities. Copper ions scattered throughout Cascariolo's piece of barite were absorbing energy when exposed to light and then slowly releasing it.

been impossible for early South American craftsmen to have melted it, though they sometimes fashioned objects from nuggets of platinum as it is relatively malleable. They also made gold-platinum sintered alloys by melting the gold and adding grains of platinum (which were found naturally as platinum-iron alloy).

Although platinum is recorded in the Old World from around 700BC, it was not recognized as distinct from silver until Spanish invaders brought it back to Europe from South America. Italian physician Julius Caesar Scaliger was the first to mention platinum in print in 1557. The name derives from the Spanish *platina*, meaning 'little silver'. This was the name indigenous miners gave to the metal, which they discarded as something they had no use for. In their search for gold, they saw platinum as a contaminant, and because of its high melting temperature could do nothing with it.

Below: Mining for platinum in South Africa, now the world's largest producer of this element.

Left: Antonio de Ulloa was an accomplished scientist and astronomer as well as an explorer and general in the Spanish navy.

William Brownrigg, back in Britain, who carried on studying it and presented his findings to William Watson, a member of the Royal Society. Details of platinum were first published in the Society's journal, *Philosophical Transactions*, in 1749 and 1750.

Platinum's exceptionally high melting point made it a challenge to work with. William Hyde Wollaston (1766–1828), an English physician who gave up his practice at the age of 35 to study chemistry, was the first to make malleable platinum. In the process, he also discovered other platinum-group metals: osmium, iridium, palladium and rhodium. Wollaston found that platinum will dissolve in *aqua regia* (see page 50), and he perfected its purification.

His decision to give up medicine, which had seemed financially risky at the time, paid off. He could buy platinum ore containing about 75 per cent platinum for 3 shillings an ounce. Then he would spend 2 shillings an ounce purifying it, and sell it on for 15 shillings an ounce. (One shilling was a twentieth of a British pound.) Wollaston made about £30,000 ($40,000) from the sale of purified platinum between 1801 and his death in 1828 (a sum equivalent to about £3 million/$4 million today). He had set out with the modest hope of making £60 ($80) a year.

Spanish naval officer Antonio de Ulloa is credited with discovering platinum. He travelled through Colombia and Peru in 1735 and found platinum as both a contaminant in gold and as whitish metal nuggets in the mines. He might have taken some back to Spain on his return in 1745. He published his observations of the metal, which he described as of 'such resistance, that it is not easy to break or cut with the force of blows on a steel anvil'. However, British chemist Charles Wood is generally credited with the discovery of platinum since he began the first thorough investigation of the metal.

In 1741, Wood found samples of platinum in Jamaica. He carried out some tests, then sent the material to the physician

Thirsty cows and magnesium

A metal that can be counted as pre-Lavoisier is magnesium, though it was not isolated for some time after its discovery. It was found in a salt, under strange circumstances.

In 1618, an English farmer called Henry Wicker had a problem with thirsty cows. During a drought, despite their thirst, his cattle refused the water he found for them. When Wicker tried the water himself he found it tasted bitter. But there was more: the water seemed to accelerate the healing of skin rashes and superficial wounds. 'Epsom salts' (magnesium sulphate) soon became famous for their therapeutic value. In 1755 Scots chemist Joseph Black realized that they contained a previously unrecognized metal, magnesium, which is reactive and difficult to replace in a compound. English chemist Humphry Davy finally extracted magnesium in 1808 using electrolysis (passing an electric current through a liquid to break down a solution, see page 93).

Trolls, gremlins and new metals

Like platinum, the next metal to be discovered had long been used in its native state, but remained unrecognized. For millennia, people had been making a brilliant blue type of pottery glaze, without knowing what imparted the vibrant colour. The element was used in Ancient China, Ancient Egypt and was found at a 3,400-year-old Danish burial site.

Above: Epsom salts have been marketed as a cure for a range of ills. Usually they are dissolved in water for bathing.

At first, chemists suspected bismuth was the key to the mystery. But Swedish chemist Georg Brandt found instead that it was cobalt. Brandt's father had been a copper smelter and Brandt grew up familiar with and interested in the chemistry of metals. Like Lavoisier later in the century, he deplored the state of chemistry at the time, and considered it little better than a pseudo-science with no solid foundations. He set about acquiring a thorough practical and theoretical education in chemistry, carrying out many careful investigations into metals and minerals. Brandt discovered cobalt while working as Guardian of the Mint in Stockholm. He became interested in a dark blue ore found in Västmanland, a local copper mining area. When he published his discovery of a new metal in 1739, other chemists contested it, claiming the ore contained a mix of iron and arsenic. Cobalt was finally confirmed as a new element by Torbern Bergman in 1780.

Cobalt ore was known to local miners as a troublesome or even cursed substance. Its name comes from the German *kobold*, meaning 'goblin'. Cobalt forms compounds with arsenic: cobaltite and skutterudite ($CoAs_3$). When an ore containing either of these is heated, it produces arsenic oxide. This could cause death to miners, or at the very least a serious sickness from which they might never recover.

It seems that mineralogy and mining were plagued with supernatural hindrances in the mid-18th century. In 1751, Swedish mining expert Axel Cronstedt discovered nickel and named it *kupfernickel*, a compound of *kupfer* (copper) and *nickel* (sprite). Nickel ore looks like copper, but miners were unable to extract any copper from it. Rather than accept that there was no copper in it, they concluded that an evil sprite was deliberately withholding it.

Just as Brandt had trouble persuading anyone that cobalt was a previously unknown material, so Cronstedt had difficulty convincing contemporary chemists that nickel was anything new. Most scientists maintained it was an alloy of cobalt (among those who accepted cobalt), arsenic, iron and copper. Again, it was Torbern Bergman who settled the case, in 1775.

Delft Blue pottery, produced in the Netherlands from the 16th century onwards, was decorated with an underglaze of cobalt blue on a white clay body, made by calcining (reducing by dessication) cobalt ore with quartz sand and potash.

Mining and metals

Mining for minerals and metal ores was the most frequent route to finding new metals. Mining had been going on for centuries – millennia, even – but we don't have much information about the procedures followed until the 16th century. Prior to that, the arcane knowledge of mining and metallurgy was passed down orally over generations. But with the invention of movable type in the late 15th century, mining knowledge began to be disseminated in written form. The most important early work on the subject was *De re metallica* by Georgius Agricola, published in Germany in 1556. (No English translation was available until 1912, when it was translated by Herbert Hoover, later 31st president of the United States.) Agricola describes (and illustrates)

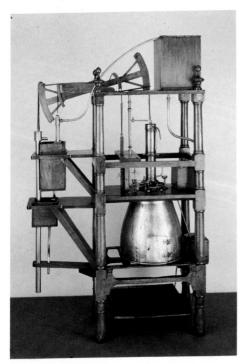

the metals being mined and the procedures and equipment for finding, extracting and preparing them, including smelting ores, separating metals, methods of assay (testing purity) and preparing the solutions needed for these processes.

Mining was about to be revolutionized with a development which, though starting in the mines, would have far wider impact. The first practically useful steam engine was invented in the early 18th century by English engineer Thomas Newcomen and installed in a mine in Cornwall between 1710 and 1714. It was used to pump water from the mine, allowing miners to go far deeper into the earth than had previously been possible. This marked the start of industrializing mining, and the first step in the Industrial Revolution. The new possibilities for mining led to the discovery of more ores and, ultimately, of more metallic elements during the 18th century.

In transition

During the late 18th and early 19th centuries, mineralogists, particularly in France, Germany and Sweden, explored the numerous ores and minerals found in mines from England to Siberia. Their processes

Left: A later version of Newcomen's innovative steam engine, this one from 1752.

'Mr Newcomen's invention of the fire engine enabled us to sink our mines to twice the depth we could formerly do by any other machinery.'

William Pryce,
Mineralogia Cornubiensis, 1778

YTTERBY'S ELEMENTS

Transition metals

p-block

d-block

| | 21 Sc | | | | | | | | | | |
| 39 Y | | | | | | | | | | | |

s-block

f-block

| | | | | | | 64 Gd | 65 Tb | 66 Dy | 67 Ho | 68 Er | 69 Tm | 70 Yb | 71 Lu |
| | | | | | | | | | | | | | |

One place proved a particularly important source of new elements in the late 18th century. The small village of Ytterby near Stockholm in Sweden has provided more new elements than any other single place on Earth, yielding ten in all: yttrium, ytterbium, terbium, erbium (all named after the village) and scandium, gadolinium, thulium, holmium, dysprosium and lutetium. There's clearly a limit to how many elements can be named after a village with only seven letters in its name.

Above: The transition, or d-block, elements form a bridge between the s-block and p-block elements. The lanthanides and actinides, or f-block elements, are sometimes considered inner transition elements.

consisted of grinding ores, heating them with carbon to remove oxygen, dissolving them in various acids and adding salts to produce precipitates (solids forced out of solution), which were then further examined. In the space of a few years, these diligent chemists discovered manganese (1774), molybdenum (1781), tellurium (1782), tungsten (1783), zirconium and uranium (1789), strontium (1790), titanium (1791), yttrium (1794), beryllium and vanadium (1797), chromium (1798), niobium (1801), tantalum (1802), and a grand haul of rhodium, palladium, cerium, osmium and iridium in 1803.

All of these, with the exception of strontium, are transition metals. In the modern Periodic Table, transition metals occupy the long band of elements lying between Group 2 (the alkaline earth metals) and Group 13 (the boron family). The transition metals in general are characterized by being hard and dense and less reactive than the alkali metals of Group 1. They form a large variety of complex ions. They were first labelled transition metals in 1921.

After 1789 and Lavoisier's new definition of a chemical element, it became possible to identify new metals not just as new but as elements. In 1807, Humphry Davy demonstrated a technique (electrolysis) for breaking apart the most stubborn of compounds and soon revealed two more very important groups of metals. Highly reactive, and with behaviour quite unlike typical metals, the alkali metals took chemistry by storm.

Wolframite is an important tungsten ore, also containing iron and manganese.

W is for 'wolf froth'

The chemical symbol for the transition metal tungsten is W, which refers to its discovery in the mineral known as wolframite. The Spanish brothers who found it, Juan José and Fausto Fermín de Elhuyar, originally named it 'wolframium', a name still used in some European countries. Confusingly, the name tungsten comes from the Swedish *tung sten*, 'heavy stone', though in Sweden the name is not used. Tung sten was the name for the mineral scheelite, later named after Carl Scheele who discovered in 1781 that a new acid could be made from it. He called this tungstic acid and suggested it might be possible to derive a metal from it.

When the Elhuyar brothers made an acid identical to tungstic acid from wolframite and went on to derive a metal from it, it was clear that this was the metal Scheele had predicted. Wolframite is a compound

---

DID IRIDIUM KILL THE DINOSAURS?

In 1803, English chemist Smithson Tennant discovered the transition metal iridium along with osmium as impurities in platinum. Osmium is the densest of the elements and iridium the second densest and the most resistant to corrosion. Iridium is extremely rare on Earth, but much more common in meteorites. The abundance of iridium in rocks laid down around 66 million years ago led to the hypothesis that the mass extinction event which killed the non-avian dinosaurs was caused by a massive asteroid strike.

of iron, manganese and tungsten oxide, and scheelite is calcium tungstate. The name 'wolframite' comes from the German *wolf rahm*, meaning 'wolf soot' or 'wolf cream', a translation of *lupi spuma* ('wolf froth'), the name coined by Agricola in 1546. Wolframite had been troublesome to German tin miners, who found its presence made tin harder to melt and generated more slag.

Centuries later, it was discovered that adding tungsten to iron made tungsten-steel. Tungsten therefore became much sought after – to the extent that even the old slag heaps of former tin mines were re-excavated in the search for it. Tungsten steel produced resilient gun barrels and gave the German army an advantage in World War I, but the supplies of tungsten stockpiled from the 1890s soon ran out, prompting a desperate search for the metal once thought a pollutant.

Tearing compounds apart

While transition elements are not difficult to extract from their mineral ores, it was harder for chemists trying to isolate the alkali metals and alkaline earth metals. They are among the most reactive of elements, and are prone to form secure compounds at the first opportunity. It's hard to prise them out of their compounds as it's difficult to find anything more reactive with which to replace them. The most reactive metals, which include potassium, sodium, lithium,

Tungsten filaments were introduced for lightbulbs in 1904. They lasted longer and glowed more brightly than the carbon filaments used previously.

barium and calcium, will react even with cold water.

The metals known in elemental form to the ancients tend to be unreactive. The least reactive metals include gold, silver, mercury, platinum and copper. They will react only with strong oxidizing acids and are extracted from ores by heating, or are simply dug from the ground in their native form. The most reactive are locked tightly into compounds and can only be separated by electrolysis.

Electrolysis followed the harnessing of electricity, which was itself a result of an observation of the strange behaviour of frogs' legs (see box on page 92). It unlocked many new elements.

FROM FROGS' LEGS TO FLUIDS

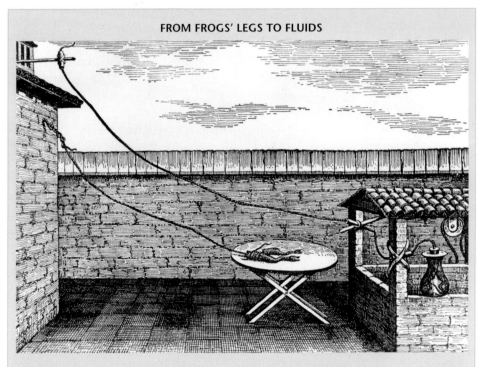

When Italian physicist and chemist Alessandro Volta invented the electric battery or 'Voltaic pile' in 1800, he opened up new possibilities in many branches of science. Volta's invention came about as a result of an argument with the Italian biologist Luigi Galvani, who had discovered that the amputated legs of frogs would generate an electric current when attached to metal wires. Galvani claimed to have discovered 'animal electricity', an animating force in living things. Volta declared that the electricity generated was produced by the metals being in contact with an electrolyte (a solution in which ions can move around). In Galvani's observation, the electrolyte was provided by fluids in the frogs' flesh. To prove his point, Volta made a similar circuit without

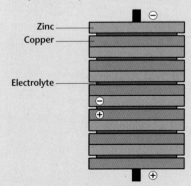

Zinc
Copper

Electrolyte

the frogs. By arranging alternating discs of copper and zinc between discs of cloth or cardboard soaked in brine (the electrolyte), he made an electric current flow through wires attached to the top and bottom of the stack. Joining the wires into a loop formed an electric circuit. He had made a world-changing discovery.

Top: One of Galvani's experiments using lightning as a source of electricity.

Left: Volta's pile or early battery.

New methods

One of those who made good early use of the battery was Humphry Davy, who began with the theory that Volta had been wrong about electricity being produced just by metals. Instead, Davy believed that a chemical reaction had produced the electric current and that, correspondingly, an electric current could prompt a chemical reaction to take place. He successfully made a Voltaic pile using zinc and carbon rather than two metals, proving his first point.

Electrolysis was the result and proved Davy's theory that passing an electric current through solutions or molten compounds causes the bonds between atoms to break apart and the components to separate. Freed positive ions then travel to the cathode (negative electrode) and negative ions travel to the anode (positive electrode).

Davy did not have the information we now have about the structure of atoms and the types of bonds they make, but he suggested that the forces which hold elements together and make them form compounds are electrical in nature. It was clear to him that the Voltaic pile and electrolysis were two facets of the same thing. He presented his work in 1807, with his discovery of potassium and sodium by 'electrical decomposition'.

Through the use of electrolysis, Davy discovered six metallic elements. All of them are metals which are not found in their native form because they are highly reactive. In the space of just two years, 1807–8, Davy discovered the alkali metals sodium and potassium, and the alkaline earth metals calcium, magnesium, strontium and barium.

ELECTRONS, ELECTROLYSIS, ELECTROLYTES, ELECTRICITY – AND IONS

An electric current involves the flow of electrons, which can carry negative electrical charge through a substance when they are released from their atoms and able to move freely. An ion is an atom or molecule that has gained or lost one or more electrons so that it has a positive or negative electrical charge. An electrolyte is a substance that forms ions in solution. Electrons will move from the negative electrode (cathode) and towards the positive electrode (anode). Salts dissolve in water, liberating ions such as potassium (K^+), sodium (Na^+), chlorine (Cl^-) and bromine (Br^-). The positive ions collect at the cathode and the negative ions at the anode, meaning that electrolysis can be used to separate a compound in solution.

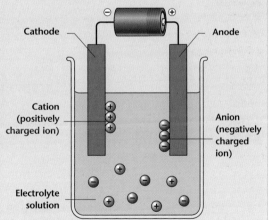

HUMPHRY DAVY (1778–1829)

Born in Penzance in the English county of Cornwall, Davy was apprenticed to an apothecary and surgeon in 1795. He began to study chemistry, and left his apprenticeship to work at the Pneumatic Institution in Bristol, where he researched gases. While there, he learned French from a refugee priest and read Lavoisier's *Traité élémentaire de chimie*. In 1798, he experimented with nitrous oxide (laughing gas), which he used recreationally as well as professionally, becoming addicted to it.

In 1801, Davy moved to London to lecture in chemistry at the Royal Institution. The next year he built the most powerful battery in the world, and used it to demonstrate the first incandescent lightbulb. He developed electrolysis as a technique and discovered several new elements with it. In 1812, he published his most important work, *The Elements of Chemical Philosophy*.

He invented the Davy lamp in 1815, in response to an approach by a group of miners from Newcastle. They complained that methane gas in mines was often ignited by the candles they had on their helmets to light their way, causing deadly explosions. Davy's wick lamp had a metal gauze sleeve protecting the flame. While gas could pass through the gauze, a flame could not. The flame would go out if the oxygen content fell below 17 per cent – a useful warning to miners.

Other public projects carried out by Davy included using chemicals to unroll carbonized papyrus scrolls retrieved from the ruins of Herculaneum, a Roman town near Pompeii that was destroyed by the eruption of the volcano Vesuvius in AD79. He also found ways to protect the copper-bottomed hulls of the Royal Navy's ships.

Scientific Researches! — New

Alkali metals and alkaline earths

Davy's first success was with the alkali metal, potassium. He first tried dissolving potash (potassium carbonate, K_2CO_3) in water, but electrolysis produced only hydrogen and oxygen. Next he put fresh, moist potash on a platinum disc, connected it to the negative pole of his battery and touched it with a platinum wire attached to the positive pole. Soon tiny globules of potassium metal began to appear on the disc. Potassium was thus the first element to be discovered by electrolysis and Davy had fun investigating its properties: '[Potassium] skimmed about excitedly with a hissing sound and soon burned with a lovely lavender light.'

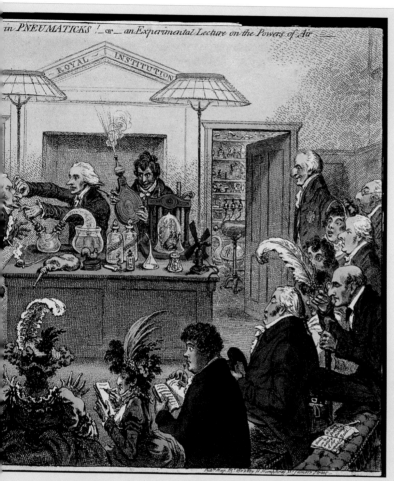

James Gillray's cartoon caricaturing demonstrations at the Royal Institution which were theatrical and popular, if not always seemly and safe!

he could not break it down – though he did suspect it might be the oxide of some unknown metal. Davy first mixed it with potash and melted the mix, then passed a current through it. It didn't work; all he got was potassium. So he tried mixing molten lime with mercury, producing an amalgam of calcium and mercury, but didn't manage to extract the alkaline earth metal, calcium. After advice from Swedish chemist Jöns Berzelius (1779–1848), he tried again, using a higher proportion of lime, and managed to make enough of the amalgam to drive off most of the mercury. He ended up with a sample of only slightly impure calcium. It would be another hundred years before a way of producing pure calcium was developed.

After his success with potassium, Davy turned his attention to caustic soda (sodium hydroxide, NaOH) and produced the alkali metal sodium from its molten form. He noted that he needed a much greater current to liberate sodium than potassium. His next target was lime (calcium oxide, CaO), which Lavoisier had listed as an element because

Davy produced barium by electrolysis of molten barium hydroxide. Barium had already been encountered in the form of barium sulphate as Bologna stone, found by Vincenzo Casciarolo in the early 1600s

Frescoes can be preserved by spraying with a solution of barium nitrate followed by ammonia. The barium hydroxide so formed fills any cracks in the plaster and slowly turns to insoluble barium carbonate.

recognized Epsom salts as the salt of an unknown metal, but had not been able to isolate magnesium. Davy had succeeded in 1808.

Strontium first came to light as a compound in 1787 when a mineral dealer in Edinburgh, Scotland, was offered a stone from a lead mine on the west coast. Thinking it was a barium mineral, he showed it to Adair Crawford, a local doctor who was exploring barium oxide's potential medicinal uses. Crawford realized it was not barium, but possibly the oxide of a new element. This was confirmed the following year by Charles Hope, another chemist in Edinburgh. In 1799, another strontium mineral, strontium sulphite, turned up. It was being used to make ornamental paths in Gloucestershire, England. Davy extracted strontium from strontium chloride in 1808.

(see page 83). Scheele had investigated it but been unable to extract the new metal he correctly suspected was present.

Magnesium had been discovered in the form of Epsom salts by Farmer Wicker's thirsty cows nearly 200 years before Davy perfected electrolysis. Joseph Black

The challenge of boron

Davy produced boron in 1808, but was beaten to winning credit for the discovery by Joseph Gay-Lussac and Louis-Jacques Thénard, who were working in Paris using the same method. They all produced it by heating borate with potassium metal, a method that had only become possible with Davy's isolation of potassium the previous year. None of them obtained a pure sample, however. Extraction of boron is extremely difficult. It's often reported as having been achieved by the American chemist Ezekiel Weintraub, who sparked a mixture of boron chloride vapour and hydrogen in 1909, but the pure polymorphs were not reported until the 1950s. The stable form of boron in ambient conditions was only found in 2007. Boron doesn't exist in its native form on Earth at all.

Both the nitride and carbide of boron remain puzzles to chemists. Boron nitride (BN), like carbon, exists in both a super-hard, sparkly form (like diamond) and a form that is soft and smooth (like graphite) but colourless. The properties depend on the shape of the crystals (cubic for the hard form and hexagonal for the soft form). Boron carbide is so hard that it is used in tank armour, but its chemical structure remains unknown.

Element overload

The proliferation of elements continued. The halogen iodine was discovered in 1811, followed by thorium, cadmium, selenium and lithium

in 1817. Although some of the new elements simply replaced compounds that Lavoisier had been unable to break down, others were entirely new. Where Lavoisier had named 33 elements (including the unlikely caloric and light), 47 had been discovered, even if not successfully isolated, by the end of 1817. It was time for someone to try to bring order to what was becoming rather chaotic. Before that happened, however, two strands of our story need to come together: the proliferation of the elements and the nature of matter. Through the work of English chemist John Dalton, Lavoisier's vision of fundamentally different types of matter would be reconciled with Boyle's corpuscularism, explained in terms of different designs of atom.

Boron carbide is almost as hard as diamond so is widely used in bullet-proof vests.

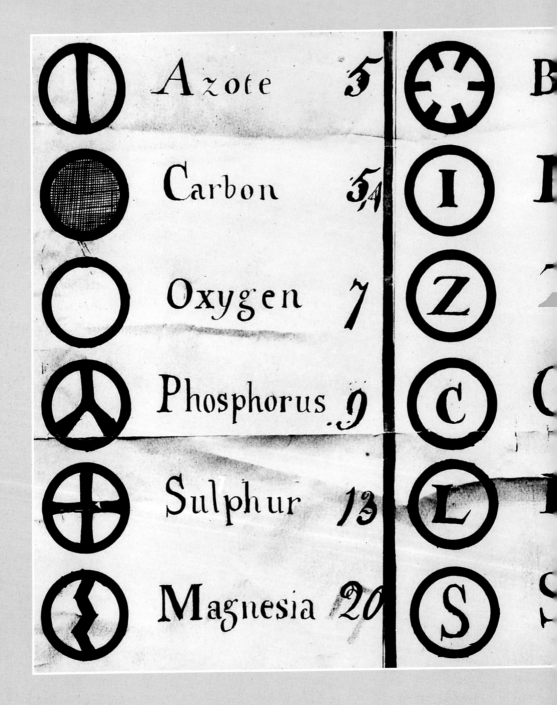

Azote	5	B
Carbon	5¼	I
Oxygen	7	Z
Phosphorus	9	C
Sulphur	13	L
Magnesia	20	S

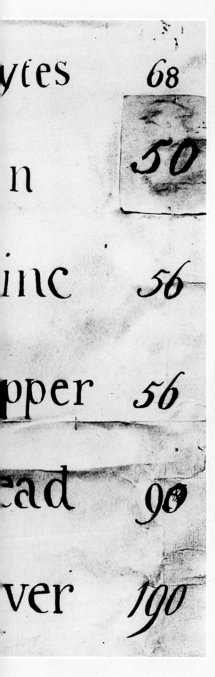

CHAPTER 5

FROM CORPUSCLES TO ATOMS

'Matter may ultimately be found to be the same in essence, differing only in the arrangement of its particles.'

Humphry Davy, 1812

The modern conception of an element relies on matter being divided into small particles that can exist and move in a void. Although the demonstration of the air pump in the 17th century helped to promote this view, it was not until the early 19th century that the existence of atoms became widely accepted.

John Dalton's revised system for depicting the elements rested on his conviction that different types of atom distinguished them.

A Renaissance for the atom

Lavoisier may have set out a new scheme for chemistry and redefined the concept of an 'element', but there was still no universally accepted model for exactly why substances differed or how they were made up. By his time, the corpuscular model was predominant. The demonstrations of the air pump and the explanation of the barometer both made a void-less model difficult to sustain. But there was much left to explain: not least, how much void there could be.

Sticking together

Boyle's version of corpuscularism explained the diversity of matter and suggested that the joining together of corpuscles could be explained by shape. If two substances didn't react, it indicated that their shapes were incompatible.

English philosopher John Locke (1632–1704) thought there was some mechanism that meant corpuscles could repel or attract

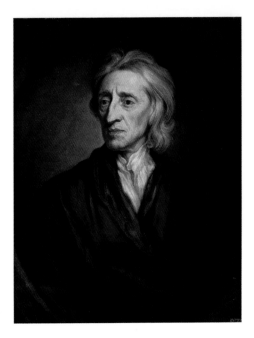

Above: John Locke had the right idea – but science had a long way to go before it could be explored.

one another, but he had no idea what it was. (Locke framed attraction and repulsion as 'connexion and repugnancy'.) He believed it would be impossible to understand chemical interactions until someone established how the mechanism worked. Locke wondered at the 'little bodies' that make up water, which are 'so extremely small' they cannot even be seen through a powerful microscope. He described how flowing water, when sufficiently cooled, freezes and 'these little atoms cohere and are not, without great force, separable.' He concluded with a wistful look at a puzzle that wouldn't be solved for several centuries: 'he that could make known the cement that makes them stick so fast one to another, would discover a great and yet unknown secret' (1690).

> 'From the various occursions of those innumerable swarms of little bodies that are moved to and fro in the world, there will be many fitted to stick to one another and so compose concretions, and many . . . disjoined from one another and agitated apart. . . . It will not be hard to conceive that there may be an incomprehensible variety of associations and textures of the minute parts of bodies, and consequently a vast multitude of portions of matter endowed with store enough of differing qualities to deserve distinct appellations.'
>
> Robert Boyle, 1661

More space, less matter

Isaac Newton adopted the model of the aggregation of corpuscles in which ever larger clusters are formed until the properties of matter and their chemical behaviour are determined. The notion that matter was made up from clusters, themselves resolvable into smaller parts, made transmutation an entirely plausible possibility. Newton was, however, certain that there remains a lot of empty space in matter. Its density depended on how closely packed the various clusters were, but even the densest materials he felt still had a lot of empty space. This was expressed spectacularly by Joseph Priestley in 1777: 'For anything we know to the contrary, all the *solid matter in the solar system* might be contained within a *nutshell*, there is so great a proportion of void.'

The prevailing model had reversed completely from the *plenum* – the packed-full universe – to suggest a universe in which there is very little, held together by forces that Newton had hinted at but which were not understood. The 'connexion and repugnancy' that Locke had noted became, in Priestley's view, the powers that rule the universe: '[the] atom must be divisible, and therefore have parts . . . [which] must

Empty space was important to Newton's idea of the universe from the micro to the macro level. In the same way as he envisaged empty space between particles, he saw the whole universe as a system of bodies with just enough space in between for their gravity to be weak enough to prevent them crashing into one another.

have powers of mutual attraction infinitely strong, or it could not hold together, that is, could not exist as a solid atom. Take away the power therefore, and the solidity of the atom entirely disappears. In short, it is then no longer matter, being destitute of the fundamental properties of such a substance.'

Priestley had been influenced by a brilliant and versatile scientist, Ruder Bošković (or Roger Boscovich, 1711–87), born in Dubrovnik. He is little known in the West today, but Bošković influenced many great scientists and mathematicians of his era. He developed a concept of the 'impenetrability' of solid bodies, based on forces rather than matter. He proposed that matter is composed of indivisible points, or atoms, which are centres of force.

True atoms

It was against this background that English chemist John Dalton introduced the basis of modern atomic theory. However, Dalton did far more than promote the atom as the foundation of matter. For the first time, he brought together the notion of the atom and the principle of the chemical elements.

Matter isn't going anywhere

Dalton's work rested on two recently formulated laws of chemistry: the law of the conservation of mass, and the law of definite proportions.

The theory of the conservation of matter had been stated by the Ancient Greek philosophers. Empedocles was adamant that nothing could be created from nothing, and so the universe had existed eternally and would always exist. When in 1756 Russian polymath Mikhail Lomonosov (1711–65)

showed that the mass of chemicals in a reaction doesn't change in a closed system, there was experimental proof that matter is not created or destroyed. The products are not created from nothing, nor are the reactants destroyed: the reactants are changed into the products entirely, with everything accounted for. Lavoisier demonstrated this more thoroughly with a series of experiments, voicing his conclusion in 1773.

The law of definite proportions was proposed by both Priestley and Lavoisier and formulated by French chemist Joseph Proust in 1794. It states that chemicals always combine in fixed proportions by weight. So, for example, Proust found that copper carbonate is always formed from copper, carbon and oxygen in exactly the same proportions. And, although tin forms two different oxides, each one is made from fixed proportions of tin and oxygen, with no intermediate versions: 100 g of tin will combine with either 13.5 g or 27 g of oxygen, but won't randomly combine completely with, say, 21 g of oxygen.

Dalton's atomic theory

Dalton suggested, first in his lectures and then in *A New System of Chemical Philosophy* in 1808, that all matter is made up of atoms of different weights. Furthermore that all atoms of a particular weight are identical and always appear in the form of the same specific substance, which is one of the elements. So, an atom of tin is not the same as an atom of oxygen or mercury, but all atoms of tin are identical, all atoms of oxygen are identical and all atoms of mercury are identical.

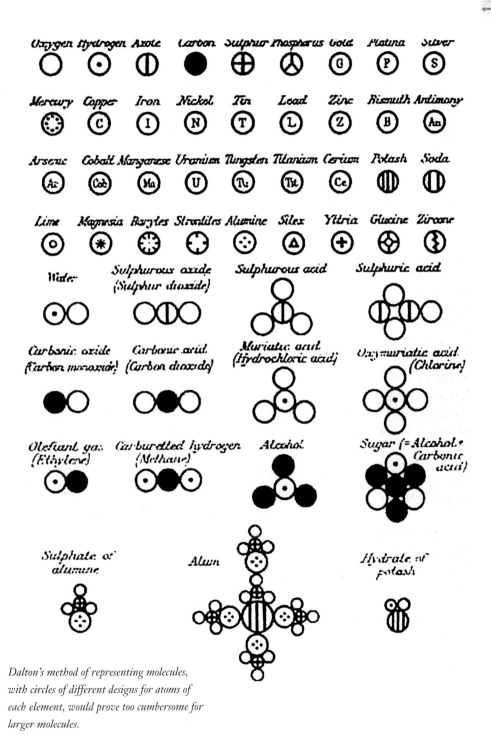

Dalton's method of representing molecules,
with circles of different designs for atoms of
each element, would prove too cumbersome for
larger molecules.

There are five parts to Dalton's atomic theory, namely:

1. All matter is made of atoms.
2. Atoms are indivisible and indestructible.
3. All atoms of an element are identical in properties and mass, and atoms of different elements are different in mass and properties.
4. Compounds are formed by combining atoms of two or more types in simple whole-number ratios.
5. A chemical reaction involves the rearrangement of atoms.

Dalton's atomic theory still largely holds true today, although points 2 and 3 have been qualified by discoveries made since his time. Atoms are no longer considered indivisible and indestructible as they are made of subatomic particles and can be broken apart by extreme force. Furthermore all atoms of an element are not necessarily identical because we now know that many elements exist as several different atomic variants (known as isotopes, see page 168). Nevertheless, Dalton's breakthrough demonstrated for the first time that matter and its behaviour could be consistently accounted for using an atomic model. His theory was in line with the conservation of matter and of mass: no atoms are destroyed

---

JOHN DALTON (1766–1844)

John Dalton was born into an impoverished Quaker household in Cumberland, England. As a religious 'dissenter' he would not have been eligible to study at university, but his background meant that he didn't have the money to go anyway. Instead he was taught by his father and a local schoolmaster, and started working as a teacher at the age of 12. From the age of 15, Dalton helped his brother run a Quaker school in the town of Kendal. In 1787, he began a weather diary which he kept for the next 57 years, making his final recording on the day before his death. His interest in meteorology led to extensive work on gases.

As a Quaker, Dalton lived a frugal life. He never married, and lived and taught in the north of England, continuing with his scientific investigations until his death. His

Drawn & Etched by J. Stephenson.

Above: Dalton was taught science by a gifted blind philosopher, John Gough.

lectures drew large audiences and were highly regarded. He was made a Fellow of the Royal Society, and of the French Académie des Sciences, and was given a state funeral on his death.

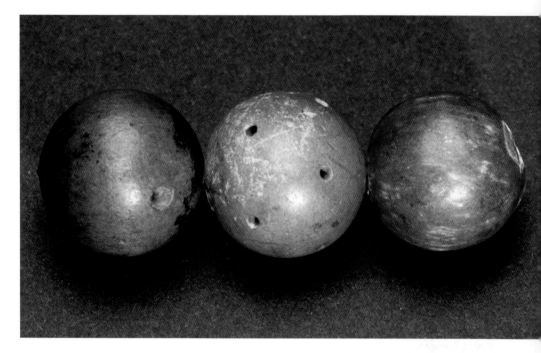

or manufactured in any process, chemical or physical, though they can be rearranged into different chemical products.

Multiple proportions and atomic weight

With point 4, Dalton extended Proust's law into the law of multiple proportions. He recognized that when atoms combine in more than one way – as, for example, when nitrogen and oxygen form nitrogen monoxide or nitrogen dioxide – the combining masses are always whole-number multiples of the smallest combining mass. In Joseph Proust's example of the oxides of tin, there will be no case in which 100 g of tin will combine with a fraction of 13.5 g or a fractional multiple of 13.5 g of oxygen. Dalton saw that these proportions must relate to the mass of the atoms involved. So

one atom of tin could combine with either one or two atoms of oxygen, but obviously it couldn't combine with fractions of an atom.

This was a brilliant insight, and it led to a century-long struggle by chemists to determine the atomic weight of all the elements. Dalton published the first list of atomic weights in 1805; it gave just six elements: hydrogen, oxygen, nitrogen, carbon, sulphur and phosphorus. Dalton took the lightest, hydrogen, as the standard, assigning it a value of 1, and worked out the comparative combining weights of the other elements.

Dalton assumed that, on the whole, atoms would combine in the ratio 1:1.

BUILDING ELEMENTS FROM HYDROGEN

In 1815, English chemist William Prout (1785–1850) noticed that none of the known atomic weights was smaller than that of hydrogen, and all seemed to be multiples of the weight of hydrogen. This led him to conclude that elements are built up from hydrogen atoms grouped together. He named this building block of the elements 'protyle'. The idea seemed plausible at first, but was overturned in 1828 when Swedish chemist Jöns Berzelius showed that not all elements have atomic weights which are whole-number multiples of the atomic weight of hydrogen. Chlorine, he found, has an atomic weight of about 35.5. This was later explained by the discovery of isotopes of chlorine with different atomic weights (see page 170), but it was sufficient to rule out protyle as the fundamental constituent of all matter. Despite this, Ernest Rutherford gave the hydrogen nucleus the name 'proton' in its honour (see page 152).

When atoms combine in only one ratio: 'It must be presumed to be a binary one, unless some cause appear to the contrary.' This conclusion led him to give the wrong formulae for some common substances, including HO for water (H_2O) and NH for ammonia (NH_3).

Insight from Italy

In the early 19th century, Britain, Germany and Sweden were the centres of chemistry. Italy was seen as a scientific backwater. Therefore, when the next significant step in understanding the atomic nature of the elements was taken in Italy by Italian count and chemist Amedeo Avogadro (1776–1856), it was ignored for a long time. Avogadro was never recognized for his

Right: Amedeo Avogadro was an Italian count and physics professor.

discovery in his lifetime, so had no impact on the chemistry of his age. Only later would his work be recognized as key to understanding and predicting how elements react together, and it would revolutionize chemistry. Like most chemists exploring the behaviour and existence of particles, Avogadro worked with gases.

There are two vitally important aspects to Avogadro's work: firstly that a fixed volume of gas contains the same number of particles, no matter which gas it is, and secondly that atoms of gas need not exist as individual entities – they can pair up into diatomic particles. This was the first time someone had suggested that a molecule might contain only one type of atom.

Building on gases

French chemist Joseph Gay-Lussac (1778–1850) carried out extensive research

The hot air balloon had been invented in 1783. Joseph Gay-Lussac was using it to sample atmospheric gases just a few years later.

into the gases which make up the atmosphere. He conducted often-dangerous trips by hot air balloon to measure pressure and humidity at high altitudes. He worked on how gases combine, and looked at the ratios in which they do so. In 1808, Gay-Lussac formulated a law which stated that 'the ratio between the volumes of reactant gases and the gaseous products can be expressed in simple whole numbers.' In other words, gases always combine in simple whole-number ratios by volume, and produce whole-number multiples of the original volumes. For example, he found that two volumes of hydrogen combine with one volume of oxygen to produce two volumes of gaseous water. He had no explanation for this, but his results were solid enough to produce a law.

Avogadro, however, could explain Gay-Lussac's law. He suggested in 1811 that if a fixed volume of gas always contains the same number of particles regardless of the gas, it's possible to add equal volumes of two different gases that will react together. For example, a litre each of gas A and gas B, will combine to produce only one litre of the gas AB, because the particles have combined into molecules. There's half the number of particles at the end, so half the volume of gas. Avogadro did not use the term atom, but 'simple molecule' (at this time, the terms 'atom' and 'molecule' were used pretty much interchangeably.)

Unpopular idea

Avogadro's idea was rejected or ignored by his contemporaries. It directly contradicted an idea of Berzelius's, that when atoms combined they were held together by some kind of electrical force, with a positive and negative attraction (a form of the 'connexion and repugnancy' proposed by Locke). It seemed this could not exist between two identical atoms of a gas: how could one oxygen atom have a positive charge and one a negative charge?

Rather than challenge the current paradigm, scientists rejected Avogadro's idea. It would lie forgotten until 1860 when another Italian chemist, Stanislao Cannizzaro, brought it to light at the Karlsruhe Congress (see page 113). This was a defining moment in the history of chemistry and the story of the Periodic Table, organized with the specific intention of sorting out the intractable mess that chemistry had become. But before the Congress, confusion would increase.

As we have seen, Avogadro realized that atoms of the same substance might come together to form molecules. This added a whole new dimension to the mathematics of chemistry as it suddenly made it possible for, say, one litre of hydrogen to contain two litres' worth of hydrogen atoms. The atoms were simply combined into one litre of hydrogen molecules for the time being, but could potentially be released from those molecules. Combining his fixed-volume insight with the realization that gases might consist of diatomic molecules, Avogadro made sense of Gay-Lussac's measurement of the combining volumes of hydrogen and oxygen to make water vapour. Gay-Lussac assumed that hydrogen and oxygen exist in atomic form, therefore:

$2H + O \rightarrow 2H_2O$ (that is, two atoms of hydrogen and one of oxygen make two molecules of water)

However, there isn't enough oxygen or hydrogen to produce this volume of water. Rewriting the equation with diatomic molecules, suggested by Avogadro, makes it work:

$2H_2 + O_2 \rightarrow 2H_2O$ (that is, four atoms of hydrogen and two of oxygen make two molecules of water)

Berzelius: positives and negatives

While Davy was using electrolysis to break down compounds in England, Berzelius was doing similar work in Sweden. He set about a series of experiments investigating a large number of chemical compounds and definitively demonstrated the law of constant proportions in inorganic compounds (that is, compounds that don't contain carbon with hydrogen). He established atomic weights for all 47 elements known at the time, which he published in 1818. (Berzelius added silicon, selenium, thorium and cerium to the haul of elements, and his students added lithium and vanadium.) His list worked not from hydrogen but from oxygen, to which he assigned a value of 100. His figures were sufficient to discredit Prout's protyle hypothesis as he showed several elements had atomic weights which are not whole-number multiples of hydrogen.

Berzelius's work led him to conclude that the forces holding atoms together in compounds are electrical. His observations of some elements gravitating towards the positive electrode and some towards the negative electrode led him to assume that the atoms had a corresponding negative or positive charge. We now know that ions, not atoms, have a charge; but ions were not recognized until the work of Michael Faraday in 1834. The nature of ionic solutions was not explained until 1884, when Swedish scientist Svante Arrhenius demonstrated that salts dissociate into ions when dissolved. Although Berzelius's theory of positively and negatively charged particles explained many inorganic couplings, it led to errors.

Representing the elements

One outcome of Berzelius's investigation was that he sought a better, more consistent and more informative way of representing the elements. The system he came up with is essentially the one we use today.

The first scheme for representing the chemical substances had been developed by the alchemists. They created different symbols for each substance; these were often taken or adapted from astrological symbols belonging to the heavenly bodies with which the substance was associated. There was no particular logic to the symbols; they simply had to be learned by rote. But they served as a shorthand for alchemists needing to write down the

Left: Berzelius introduced the modern method for representing the elements, using one or two letters from their names in English or Latin.

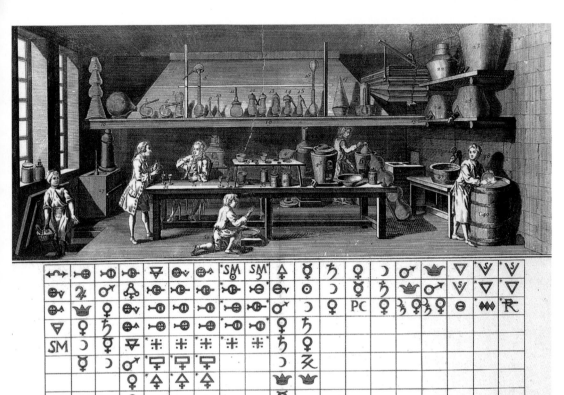

Above: The alchemists' method of a graphic symbol for each element was challenged by the discovery of new elements.

name of a chemical and conveniently helped to keep their writings obscure and secret. This was a great improvement over the alchemists' names such as 'butter of antimony' (antimony chloride), and 'lunar caustic' (silver nitrate).

Lavoisier introduced method to the previous madness of the naming of chemicals. Having established that everything is made of elements, it seemed logical to name chemicals according to their constituents.

John Dalton, perhaps influenced by alchemical tradition, returned to a system of graphic symbols. He represented atoms by circles, with a different design of circle for each element. At first glance, it looks as though some are easy to guess, as they show a letter inside a circle. But it's not as easy as it looks: P is not phosphorus but platinum, C not carbon but copper, and S not sulphur but silver. Unhelpfully, his symbol for hydrogen was the same as the symbol the alchemists had used for gold.

Molecules were represented by putting the atoms together. Given that

110

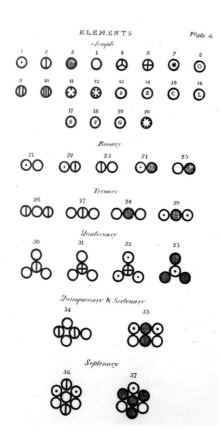

Dalton erroneously thought that water comprised one atom of hydrogen and one of oxygen and that ammonia was one atom of hydrogen with nitrogen, he showed water and ammonia as ⊙◯ and ◯◐ respectively.

Adding more atoms made more complex designs that are not easy to read at a glance: as the number of symbols proliferated, remembering them became more burdensome. There were other problems with the method, too.

Berzelius developed a better system which he publicized in 1813–14. Instead of special symbols, he used one or two letters of the element's Latin name. Latin was the lingua franca of science at the time, so the scheme was as international as possible. Dalton's system of circles, with symbols such as G in a circle for gold and S in a circle for silver, was distinctly anglophone.

Left: Dalton's system for representing the elements stuck circles together to show compounds.

'The chemical signs ought to be letters, for the greater facility of writing, and not to disfigure a printed book. Though this last circumstance may not appear of any great importance, it ought to be avoided whenever it can be done. I shall take, therefore, for the chemical sign, the initial letter of the Latin name of each elementary substance: but as several have the same initial letter, I shall distinguish them in the following manner:–

1. In the class which I call metalloids, I shall employ the initial letter only, even when this letter is common to the metalloid and some metal.

2. In the class of metals, I shall distinguish those that have the same initials with another metal, or a metalloid, by writing the first two letters of the word.

3. If the first two letters be common to two metals, I shall, in that case, add to the initial letter the first consonant which they have not in common: for example, S = sulphur, Si = silicium, St = stibium (antimony), Sn = stannum (tin), C = carbonicum, Co = cobaltum (cobalt), Cu = cuprum (copper), O = oxygen, Os = osmium, &c.'

Jöns Berzelius, 1813–14

Berzelius's system was also infinitely extendable. As more elements were discovered and named, just a couple of letters from their name provided the symbol. If the obvious letter or pair had already been used for another element, a letter from further on in the word was used for the new element; hence, Hg for mercury (*hydrargyrum*, meaning 'water silver') and Pb for lead (*plumbum*).

The formula for a molecule was shown by adding a superscript (above the line) numeral if there was more than one atom of a particular element in a molecule, as in 'H²O'. Current usage deviates from Berzelius's original plan, as the numbers are now subscript (below the line), as in H_2O. But, either way, this indicates that each water molecule consists of two atoms of hydrogen and one of oxygen.

Chemistry puts its house in order

Most chemists accepted Dalton's atomic theory and the fact that both atoms and molecules exist, but there was considerable disagreement about atomic weights. Should oxygen be 8 or 16? Should carbon be 6 or 12? They also could not agree about the

> 'More precise definition of what is meant by the expressions: atom, molecule, equivalence, atomicity, basicity, and designated expressions; investigation as to the true equivalent of bodies and their formulas; introduction of a proportional description and a rational nomenclature.'
>
> Goals of the Karlsruhe Congress, as stated in the invitation of 10 July 1860

$C_4H_4O_4$	empirische Formel.
$C_4H_3O_3 + HO$	dualistische Formel.
$C_4H_3O_4 \cdot H$	Wasserstoffsäure-Theorie.
$C_4H_4 + O_4$	Kerntheorie.
$C_4H_3O_2 + HO_2$	Longchamp's Ansicht.
$C_4H + H_3O_4$	Graham's Ansicht.
$C_4H_3O_2 \cdot O + HO$	Radicaltheorie
$C_4H_3 \cdot O_3 + HO$	Radicaltheorie.
$\left.\begin{smallmatrix}C_4H_3O_2\\H\end{smallmatrix}\right\}O_2$	Gerhardt. Typentheorie.
$\left.\begin{smallmatrix}C_4H_3\\H\end{smallmatrix}\right\}O_4$	Typentheorie(Schischkoff)etc.
$C_2O_3 + C_2H_3 + HO$	Berzelius' Paarlingstheorie.
$HO \cdot (C_2H_3)C_2, O_3$	Kolbe's Ansicht.
$HO \cdot (C_2H_3)C_2, O \cdot O_2$	ditto
$\left.\begin{smallmatrix}C_2(C_2H_3)O_2\\H\end{smallmatrix}\right\}O_2$	Wurtz.
$\left.\begin{smallmatrix}C_2H_3(C_2O_2)\\H\end{smallmatrix}\right\}O_2$	Mendius.
$\left.\begin{smallmatrix}C_2H_2 \cdot HO\\HO\end{smallmatrix}\right\}C_2O_2$	Geuther.
$C_2\left\{\begin{smallmatrix}C_2H_3\\O\\O\end{smallmatrix}\right.O + HO$	Rochleder.
$\left(C_2\frac{H_3}{CO} + CO_2\right) + HO$	Persoz.
$C_2\left\{\begin{smallmatrix}C_2\{O_2\\H\\\hline H\\\hline H\end{smallmatrix}\right.\}O_2$	Buff.

Above: The list of possible formulae for acetic acid shows the extent to which chemistry was floundering in the 19th century. There was no way of knowing how atoms might be arranged in a molecule.

existence of diatomic molecules. This meant they could not even reach a consensus about the formula of water or describing the reactions that produce it. The configuration of molecules was hotly disputed; at least 19 different ways of representing acetic acid were in circulation!

Watershed at Karlsruhe

It fell to a young German chemist, August Kekulé, to take up the challenge. Kekulé had already proposed (correctly) that

STANISLAO CANNIZZARO (1826–1910)

Stanislao Cannizzaro combined innovative chemistry with military service, revolutionary fervour and a passion for politics. At the age of 23, following the failure of the Sicilian war of independence during which he had been an artillery officer, he fled to Marseille, France. Two years later, he returned to Italy to become professor of chemistry at Genoa. He later tried to join another rebellion in Sicily, arriving too late to take part in the fighting but in time to join the new government. As well as the Karlsruhe paper, he is known in chemistry for 'Cannizzaro's principle' (that the mass of one atom of an element is the smallest mass of that element found in any molecule containing it) and 'Cannizzaro's reaction' (in which benzaldehyde is combined with potassium carbonate to form benzyl alcohol and potassium benzoate). For his achievements, the Royal Society awarded him the Copley Medal in 1891.

Stanislao Cannizzaro

carbon can form four bonds. He would go on to suggest the structure of the benzene ring in 1865. In 1860, in the German city of Karlsruhe, Kekulé brought together the leading chemists of Europe to thrash out an agreement and consistent basis for their discipline. The Karlsruhe Congress was the first international science conference.

As well as the great figures of Europe, 127 chemists from as far afield as Russia and Mexico attended the Congress. They included some big names: Bunsen, Meyer and Borodin (a chemist as well as a composer). Most papers were presented by chemists with traditional views, but at the end a young Sicilian chemist, Stanislao Cannizzaro (see box), delivered the final paper – an eloquent and enthusiastic defence of Avogadro's conception of molecules and atoms. Four years after Avogadro's death, his groundbreaking theory was finally given a sympathetic hearing.

Despite Cannizzaro's astonishing paper, the Congress made little immediate impact, but a seed had been planted. Lothar Meyer (1830–95), who before the Congress had referred to it as an 'idiotic church-council', wrote afterwards that, 'It was as though the scales fell from my eyes; doubt vanished, and it was replaced by a feeling of peaceful certainty.'

One of those present at Karlsruhe was a young Russian, Dmitri Mendeléev, who taught chemistry for a living. Soon afterwards, he would take significant steps to help bring order to the chemical chaos.

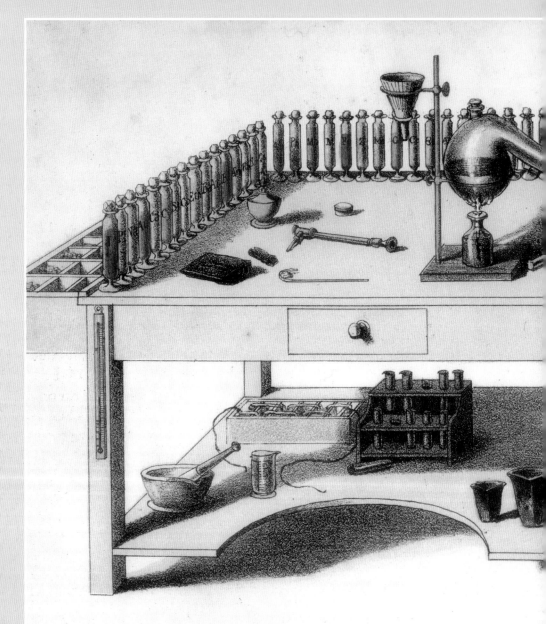

THE LABORATORY AT HOME.

CHAPTER 6

THE SEARCH FOR ORDER

'It is the function of science to discover the existence of a general reign of order in nature and to find the causes governing this order.'

Dmitri Mendeléev, 1901

The increasing number of elements emerging from the work of 19th-century chemists was matched by an increasing alarm at their abundance. The world had once been neatly contained in just four elements, but now it seemed that there were too many to grasp. Chemists began to search for some expression of natural order in the proliferating elements.

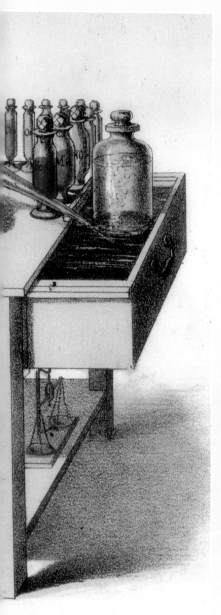

M.& N.Hanhart

A chemist's bench in the 1860s, of the type that an amateur scientist might be able to put together at home.

Three by three, four by four

In 1817, German chemist Johann Döbereiner (1780–1849) made the first attempt to find order in the elements. Noticing that the newly discovered bromine seemed to have properties similar to those of chlorine and iodine and that its atomic weight was approximately midway between the two, he suggested that they form some kind of triad. He found two other triads: calcium, strontium and barium; and sulphur, selenium and tellurium. He wasn't able to find more, though, so his system, which he called the Model of Triads, did not appear to be particularly useful. We can now see that he'd identified a set of halogens: chlorine, iodine and bromine; a set of alkaline earth metals: calcium, strontium and barium; and part of Group 16, with sulphur, selenium and tellurium. There was potential to extend the 'triads' further, as the two elements above calcium, strontium and barium – beryllium and magnesium – were also known. But if Döbereiner noticed similarities, he made nothing of them.

That step was taken by Jean-Baptiste Dumas. In 1859, he extended the triads to groups of four, adding fluorine to Döbereiner's three halogens (so fluorine, chlorine, bromine and iodine) and magnesium to the alkaline earths (giving magnesium, calcium, strontium and barium). Again, he couldn't find other groups of four, which might have suggested that it was the basis of a truly useful system of classification. But Döbereiner's insight that atomic weight might have something to do with patterning in the elements had set other chemists looking in roughly the right direction.

Atomic weights revisited

In the early 19th century, atomic weights had not been reliably established; in some cases, chemists were working with the wrong weights when trying to detect periodicity. The concept of atomic weight only emerged with

The halogens include elements in all three states of matter at room temperature. Left to right: chlorine is a green gas, bromine a brown liquid and iodine a volatile purple solid.

Right: Crooke's 'Vis generatrix' is a 3D model that tries to visualize the relationship between the elements. Similar elements are arranged vertically above one another, with the element of lowest atomic weight at the bottom. White tokens indicate a missing element.

Dalton and his conviction that atoms, though tiny, are solid and have mass. The early method for determining atomic weight was to compare the masses of gases that combine together, starting with hydrogen as the standard. Dalton assumed that oxygen and hydrogen combined in the ratio 1:1 to make water, so took the atomic weight of oxygen to be 8. His assumption that ammonia is one part nitrogen to one part hydrogen gave him an atomic weight of 5 for nitrogen . . . and so on.

Berzelius revisited atomic weight. The results of his experiments, published in 1828, are not far from the modern values. He corrected Dalton's errors and extended the list accurately, largely as a result of the work of two French physicists, Pierre Dulong and Alexis Petit, published in 1819. They had derived the Dulong–Petit law of thermodynamics, which states that the heat capacity of a certain weight of an element multiplied by its atomic weight is a constant. Once this had been established, it was easy to turn it around – starting with the constant and a measured heat capacity – to

work out the atomic weight of an element (or the molecular weight of a compound). This meant that, after 1828, more accurate values for atomic weights were available – though not everyone agreed on them.

THE MARCH OF THE ELEMENTS

Between Dalton's publication of *New System of Chemical Philosophy* in 1808 and the Karlsruhe Congress, 14 further elements were discovered. These were: iodine (1811), lithium (1817), cadmium (1817), selenium (1818), silicon (1823), aluminium (1825), bromine (1826), thorium (1828), vanadium (1830), cerium (1839), lanthanum (1839), terbium (1843), erbium (1843) and ruthenium (1844).

The Telluric Screw

In 1863, French geologist Alexandre-Émile Béguyer de Chancourtois tried to show the relationships between elements geometrically. He developed a system he called the Telluric Screw (because tellurium fell in the middle of the geometric pattern). He derived his display of periodicity by writing down the elements in order of increasing atomic weight on a long piece of paper and wrapping the list around a cylinder in such a way that those with similar properties (the previously noted groups of three and four) lined up with one another. One turn of the cylinder was completed at every sixteenth element. Although some elements were in the right place, others were not, so it was not an accurate system. And even though there were hints of some kind of repeating pattern if the elements were correctly ordered, there was nothing sufficiently consistent to withstand much scrutiny and no hint as to how or why properties should follow any kind of pattern.

Musical elements?

Four years after the Karlsruhe Congress, two chemists proposed repeating patterns in the elements when arranged in order of atomic weight. William Odling suggested that similar properties repeat at intervals of seven, and John Newlands proposed the Law of Octaves, again arranging the elements in rows of seven with the eighth, he said, resembling the first. Although the system worked for two rows, it failed after the element calcium. It is here that the current Periodic Table launches into the first long row of transition metals, which appear before the Group 13 elements.

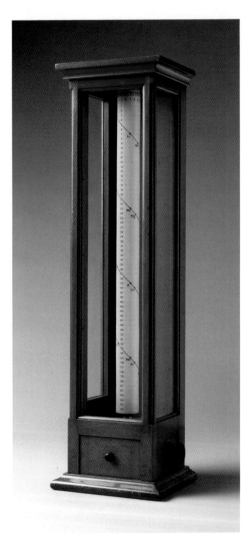

Above: The Telluric Screw, showing the paper with the list of elements wound around a cylinder.

Mendeléev cracks it

The person credited with putting together the Periodic Table in the form we now know it is Dmitri Mendeléev, a teacher and author of chemistry textbooks. While writing the definitive textbook of the time,

Principles of Chemistry, he solved the issue of periodicity. The book was the reason he had to think about the issue, as he was treating collections of elements sequentially. He had written about the alkali metals in the first part of the book, and needed to decide which elements most closely resembled them in order to write the next part. Fortunately for historians of science, Mendeléev was obsessive about keeping his documentation, preserving even the most apparently useless scraps of paper.

Mendeléev began by comparing the alkali metals with other groups of metals, but focusing on differences in atomic weight. Chemists already loosely grouped elements on the basis of similar properties. The alkali metals are all highly reactive, reacting even with water; the halogens all form salts when reacted with the alkali metals. Like the early classification of plants and animals by shape and behaviour, this groups elements by their properties rather than innate

structure. Since many properties are the result of innate structure, it works – up to a point. Mendeléev suspected that the properties of the elements were the product of something fundamental.

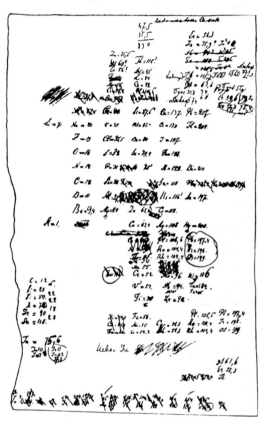

Right: Mendeléev's early notes on the arrangement of the Periodic Table, 1869.

A MOMENT OF CLARITY

Mendeléev's magpie tendency led to the moment of his inspiration being preserved. In the Mendeléev Archive in St Petersburg, chemist and researcher Bonifaty Kedrov discovered a document recording Mendeléev's insight into the problem of the Periodic Table. He was interested in agricultural reform and the new economic models that were emerging in agrarian Russia and he had arranged to visit a dairy co-operative. On 17th February 1869, while reading a letter setting out the schedule for his visit the next day, he scribbled down on the back of it his first experimental groupings of what would become the Periodic Table.

Mendeléev's first efforts were not very enlightening, but the idea that atomic weights were the clue stuck with him. He noted down the non-metallic elements with their atomic weights, putting the group of halogens in a horizontal row and placing the elements with the closest atomic weights in rows beneath them. Mendeléev's extensive work as a practical chemist meant he was familiar with how the elements would behave and react; the trick was to identify the features that were important for grouping them.

He managed to arrange 42 of the 63 elements then known. The others were

Below: Mendeléev's first Periodic Table.

> '[Mendeléev was] a peculiar foreigner, every hair of whose head acted independently of every other.'
> Scottish chemist Sir William Ramsay, on meeting Mendeléev in London

more problematic as less was known about their properties and even their atomic weights were not certain. The work was slow and frustrating, as Mendeléev had to copy out his table afresh each time he wanted to try an element in a new position.

Mendeléev was an enthusiastic player of the card game 'Patience' and that, ultimately, provided the breakthrough. It occurred to him, as he struggled despondently over his repeatedly copied tables, that he could write the name, atomic weight and characteristics of each element on a separate card and then move the cards around at will. He had by this stage come to visualize the form of the answer he was seeking: a table in which elements with similar features were arranged into groups horizontally, and elements with similar atomic weights were arranged vertically. (This is the opposite arrangement to the Periodic Table we have now, with atomic numbers and weights in horizontal sequence.) The principle was sound, but the difficulty lay in finding the correct location for each card. Mendeléev had long felt that there should be some relation

ОПЫТЪ СИСТЕМЫ ЭЛЕМЕНТОВЪ.

ОСНОВАННОЙ НА ИХЪ АТОМНОМЪ ВѢСѢ И ХИМИЧЕСКОМЪ СХОДСТВѢ.

		Ti = 50	Zr = 90	? = 180.
		V = 51	Nb = 94	Ta = 182.
		Cr = 52	Mo = 96	W = 186.
		Mn = 55	Rh = 104,4	Pt = 197,4.
		Fe = 56	Rn = 104,4	Ir = 198.
		Ni = Co = 59	Pl = 106,6	O- = 199.
H = 1		Cu = 63,4	Ag = 108	Hg = 200.
Be = 9,4	Mg = 24	Zn = 65,2	Cd = 112	
B = 11	Al = 27,4	? = 68	Ur = 116	Au = 197?
C = 12	Si = 28	? = 70	Sn = 118	
N = 14	P = 31	As = 75	Sb = 122	Bi = 210?
O = 16	S = 32	Se = 79,4	Te = 128?	
F = 19	Cl = 35,5	Br = 80	I = 127	
Li = 7	Na = 23	K = 39	Rb = 85,4	Cs = 133	Tl = 204.
		Ca = 40	Sr = 87,6	Ba = 137	Pb = 207.
		? = 45	Ce = 92	
		?Er = 56	La = 94	
		?Yt = 60	Di = 95	
		?In = 75,6	Th = 118?	

Д. Менделѣевъ

ПЕРИОДИЧЕСКАЯ СИСТЕМА ЭЛЕМЕНТОВ

ГРУППЫ ЭЛЕМЕНТОВ

ПЕРИОДЫ	РЯДЫ	I	II	III	IV	V	VI	VII	VIII	0
1	I	H 1 1,008								He 2 4,003
2	II	Li 3 6,940	Be 4 9,02	B 5 10,82	C 6 12,010	N 7 14,008	O 8 16,000	F 9 19,00		Ne 10 20,183
3	III	Na 11 22,997	Mg 12 24,32	Al 13 26,97	Si 14 28,06	P 15 30,98	S 16 32,06	Cl 17 35,457		Ar 18 39,944
4	IV	K 19 39,096	Ca 20 40,08	Sc 21 45,10	Ti 22 47,90	V 23 50,95	Cr 24 52,01	Mn 25 54,93	Fe 26 55,85 / Co 27 58,94 / Ni 28 58,69	
4	V	Cu 29 63,57	Zn 30 65,38	Ga 31 69,72	Ge 32 72,60	As 33 74,91	Se 34 78,96	Br 35 79,916		Kr 36 83,7
5	VI	Rb 37 85,48	Sr 38 87,63	Y 39 88,92	Zr 40 91,22	Nb 41 92,91	Mo 42 95,95	Ma 43 —	Ru 44 101,7 / Rh 45 102,91 / Pd 46 106,7	
5	VII	Ag 47 107,88	Cd 48 112,41	In 49 114,76	Sn 50 118,70	Sb 51 121,76	Te 52 127,61	J 53 126,92		Xe 54 131,3
6	VIII	Cs 55 132,91	Ba 56 137,36	La 57 138,92 ★	Hf 72 178,6	Ta 73 180,88	W 74 183,92	Re 75 186,31	Os 76 190,2 / Ir 77 193,1 / Pt 78 195,23	
6	IX	Au 79 197,2	Hg 80 200,61	Tl 81 204,39	Pb 82 207,21	Bi 83 209,00	Po 84 210	85 —		Rn 86 222
7	X	87 —	Ra 88 226,05	Ac 89 227	Th 90 232,12	Pa 91 231	U 92 238,07			

★ ЛАНТАНИДЫ 58–71

Ce 58 140,13	Pr 59 140,92	Nd 60 144,27	61 —	Sm 62 150,43	Eu 63 152,0	Gd 64 156,9
Tb 65 159,2	Dy 66 162,46	Ho 67 164,94	Er 68 167,2	Tu 69 169,4	Yb 70 173,04	Cp 71 174,99

between the atomic weight and properties of elements, if only it could be discovered. He occasionally noted the layout of his cards and symbols of elements he had yet to place, crossing them off as he incorporated them. These preserved pieces of paper give a full insight into his processes. By the end of the first day, Mendéleev had found places in his table for 56 elements, with seven he could not place. These would, he realized, take further investigation.

Out of order

Mendéleev was not the first to use atomic weight as the organizing principle in trying to sort out the elements. But he was the first to be guided by both atomic weight and a deep personal knowledge of the properties of the elements. Crucially, he let the second take precedence when there was a conflict. This meant that occasionally he would allow elements to switch places so they were in the right place according to their properties, even though it meant their atomic weights were out of sequence. For example, he moved beryllium (atomic weight 14) from its position above nitrogen and placed it above

Above: Mendéleev's Periodic Table re-arranged as in the modern table. This arrangement goes straight from Group 2 (starting with beryllium) to Group 13 (starting with boron). The elements now classified as transition elements are show to the right, before the noble gases (mostly unknown until the 1890s).

DMITRI MENDELÉEV (1834–1907)

One of about 17 children, 14 of whom died in infancy, Mendeléev was born in Siberia and raised an Orthodox Christian, although he later left the church. His father went blind and had to give up his teaching job when Mendeléev was a young child. His mother restarted a glass factory that was in her family but had been abandoned. Despite her obvious resilience, the family still did not thrive. When Mendeléev was 13, his father died and the glass factory was destroyed by fire. His mother took him to Moscow, where the university refused to give him a place. From there, mother and son went on to St Petersburg where in 1850 Mendeléev was admitted to the Main Pedagogical Institute (now St Petersburg State

Mendeléev deliberately cultivated his wild, mad-scientist look, having his hair cut by a sheep-shearer just once a year.

University). Ten days after his admission, Mendeléev's mother died of tuberculosis. Soon after graduating, he contracted tuberculosis himself.

Still he persisted. He worked at the University of Heidelberg with German chemist Robert Bunsen (1811–99), inventor of the Bunsen burner, and attended the Karlsruhe Congress in 1860. When he returned to Russia in 1861, he taught chemistry in St Petersburg, becoming professor at the university at the age of 33. Dismayed at the poor quality of textbooks available to his students, Mendeléev set out to write his own, focusing on organic and then inorganic chemistry. They became the pre-eminent texts of the time, used into the 20th century, even outside Russia.

Mendeléev's development of the Periodic Table did not bring overnight fame, but once the first of his predicted elements, gallium, was discovered his reputation was secure. Not only a superlative chemist, Mendeléev was also a skilled maker of luggage and had interests in shipbuilding, petroleum and agriculture. He was involved in the construction of the first Arctic icebreaker, the *Yermak*, launched in 1898. Despite his reputation, Mendeléev was also capable of making mistakes. He proposed that hydrocarbons such as petroleum come from deep in the Earth where they were formed by inorganic means, and he proposed two inert elements, lighter than hydrogen, which he thought formed the chemical composition of *aether*.

I	II	III	IV	V	VI	VII	VIII		
H 1.01									
Li 6.94	Be 9.01	B 10.8	C 12.0	N 14.0	O 16.0	F 19.0			
Na 23.0	Mg 24.3	Al 27.0	Si 28.1	P 31.0	S 32.1	Cl 35.5			
K 39.1	Ca 40.1		Ti 47.9	V 50.9	Cr 52.0	Mn 54.9	Fe 55.9	Co 58.9	Ni 58.7
Cu 63.5	Zn 65.4			As 74.9	Se 79.0	Br 79.9			
Rb 85.5	Sr 87.6	Y 88.9	Zr 91.2	Nb 92.9	Mo 95.9		Ru 101	Rh 103	Pd 106
Ag 108	Cd 112	In 115	Sn 119	Sb 122	Te 128	I 127			
Ce 133	Ba 137	La 139		Ta 181	W 184		Os 194	Ir 192	Pt 195
Au 197	Hg 201	Ti 204	Pb 207	Bi 209					
			Th 232		U 238				

Above: The Periodic Table as Mendeléev left it. There is no dedicated space for the transition metals (see page 89), some of which have been crammed into other groups, no group for the noble gases and no final period (the bottom row in the modern Periodic Table, starting with francium).

magnesium in Group 2. This was certainly where it belonged in terms of its properties, but it would need an atomic weight of 9 to be in sequence.

By positioning the elements according to their properties rather than their atomic weights, Mendeléev reduced the detrimental impact of the greatest limitation facing chemists working to organize the elements in the 19th century: their dependence on atomic weight rather than the as yet unknown atomic numbers. As we shall see, once that hurdle was overcome in the 20th century, everything fell into place.

Not quite first

Although Mendeléev is credited with organizing the elements by atomic weight, German chemist Lothar Meyer published a similar, smaller table in 1864. He plotted atomic weight against atomic volume and achieved a series of peaks and troughs, with the alkali metals at the peaks and the halogens at the troughs. (This is now recognizable as descending electropositivity: an element's readiness to lose electrons, decreasing from the alkali metals to the halogens.) Using this as the basis of periodicity, he drew up a table of 28 elements. He considered the relative advantages and disadvantages of both horizontal and vertical arrangements.

> 'To search for something – though it be mushrooms – or some pattern – is impossible, unless you look and try.'
>
> Dmitri Mendeléev

Mind the gap

Mendeléev left a space for elements he expected would be discovered in future and even made predictions about their properties and atomic weight. He predicted eka-aluminum (gallium, Ga), eka-boron (scandium, Sc), and eka-silicon (germanium, Ge). His prefix 'eka' was from Sanskrit, meaning 'first', and it was added to the name of the element above the gap. (He had further prefixes, dvi- (2) and tri- (3) for predicting subsequent unknown elements in a group.)

Gallium was identified in 1875 by French chemist Paul Emile Lecoq de Boisbaudran; scandium was found in 1879; and germanium in 1886. In a flush of patriotic pride, de Boisbaudran named gallium after the Latin word for 'France'. Most of its properties closely matched Mendeléev's predictions for eka-aluminium; however, de Boisbaudran's value for the density of gallium was lower than Mendeléev had predicted. Mendeléev wrote to de Boisbaudran, informing him that he had already predicted the properties of the metal and pointing out de Boisbaudran's error. When de Boisbaudran double checked, he found that the true value was closer to Mendeléev's prediction.

Ghost discoveries

Mendeléev's predictions inevitably led to chemists rushing to find his missing, predicted elements, and naming them themselves. This resulted in an avalanche of renaming. For example, Mendeléev's predicted element eka-caesium eventually turned out to be francium, discovered in 1939 by Marguerite Perey as a product of the radioactive decay of actinium-227. In between Mendeléev's prediction and Perey's discovery, eka-caesium was 'found' four times. The first time by Soviet chemist D.K. Dobroserdov in 1925, who named it 'russium' after his home country. Then, in 1926, two English chemists, Gerald Druce

Predicted	Actual	Discovered
Eka-boron	Scandium	1879
Eka-aluminium	Gallium	1875
Eka-manganese	Technetium	1937
Eka-silicon	Germanium	1886
Tri-manganese	Rhenium	1925
Dvi-tellurium	Polonium	1898
Dvi-caesium	Francium	1939
Eka-tantalum	Protactinium*	1900

Left: The elements predicted by Mendeléev, showing his names for them, their final names and the dates they were found.

* The place assigned to be eka-tantalum, between thorium and uranium, was eventually occupied by dubnium, as protactinium is classed as an actinide and therefore is out of the sequence of transition metals.

and Frederick Loring, found the spectral lines which they thought belonged to eka-caesium using a new technique called spectroscopy (see below). They named it 'alkalinium', as it would be the heaviest alkali metal. In 1930, eka-caesium was 'found' in the USA by Fred Allison, who wanted to call it 'virginium' after his home state of Virginia. Then Romanian and French physicists Horia Hulubei and Yvette Cauchois believed they had found the emission spectrum of the element in 1936, proposing the name 'moldavium' after the Romanian province of Hulubei's birth. Chemists were hunting for several missing elements, and francium was far from unique in having many ghost discoveries before it was finally found.

No room for gas

In addition to the gaps Mendeléev reserved for elements he knew would be discovered, there is another glaring omission in his table, between the halogens and the alkali metals. As none of the noble gases was known at the time, he had no reason to suspect that anything would lie between his groups I and VIII. In fact, helium had been discovered in 1868, the year before Mendeléev's work

on the Periodic Table, but no one believed in it. That – and the landscape of chemistry – would change with the work of English chemist Sir William Ramsay (1852–1916). The discovery of helium, and its final confirmation as a new element, rested on a technique that has become extremely important in many areas of chemistry, not least in the identification of elements.

Below: The atomic structure of francium, element 87, which was 'found' four times before being finally discovered in 1939, 70 years after Mendeléev first predicted its existence.

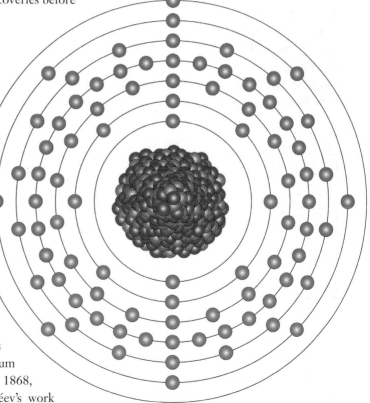

Looking into the light

The identification of helium was made possible by the discovery, in 1859, of the spectroscopic 'fingerprint' of the elements by eminent German chemist Robert Bunsen and physicist Gustav Kirchhoff (1824–87). Their work rested on a series of discoveries, beginning with the work of Isaac Newton.

Lines in the light and in the dark

It's clear from observing a rainbow that the apparently white light of the Sun hides a multitude of colours. In 1666, Newton demonstrated that light of different colours makes up white light and is differently refracted (forced to change direction). He split white light into the spectrum by passing it through a glass prism and reconstituted the white light using a second prism.

Nearly 100 years later, in 1752, Scottish physicist Thomas Melvill found that if he put different substances into flames, then passed the light of the flame through a glass prism, the result was a 'broken' spectrum that varied from one substance to another. Unlike the splitting of sunlight, the light from a flame produced only a partial rainbow of colours with dark gaps in the spectrum. In some cases the result was largely darkness with just a few bands of colour. The next century, even the spectrum of sunlight would be found to be broken.

In 1814, a talented German lens-maker, Joseph von Fraunhofer, took Newton's discovery a step further and found that by bringing the light to the prism through a very narrow slit, he could separate the components of sunlight still further, revealing 574 dark bands in the spectrum. Fraunhofer plotted the dark lines carefully, but their origin eluded him, and several investigators who came after him.

A few years later, British scientists John Herschel and William Fox Talbot (both pioneers of photography) suggested that the distinctive colour of a flame might be used to indicate the presence of particular elements in a substance. Fox Talbot noted the potential of spectroscopy in chemical analysis in 1826: 'I would further suggest that whenever the prism shows a homogenous ray of any colour, to exist in a flame, this ray indicates the formation or the presence of a definite chemical compound.'

But Fox Talbot didn't pursue it because of an unsettling inconsistency. Although he recognized the yellow light characteristic of sodium, he often found it where no sodium was present, so distrusted the usefulness

Below: Fraunhofer was first to find the dark bands in the spectrum of sunlight. Now called Fraunhofer lines, their position indicates the absorption of light of specific wavelengths by gases in the atmosphere of a star.

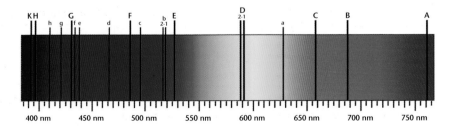

Above: Spectral data is used to produce false-colour images of celestial bodies (Mercury, in this case) that indicate the composition of the surface.

of the method. In fact it was the purity of his samples rather than the robustness of spectroscopy that was in doubt. Nevertheless Fox Talbot and Herschel were right about the potential of spectroscopy: it is now a popular method for investigating the composition of matter, particularly useful for identifying the distant matter of stars and even other galaxies. Yet it would be 25 years more before its potential began to be explored rigorously.

Looking for lines

In 1851, Bunsen and Kirchhoff were working together at the University of Heidelberg in Germany. Bunsen had become so fascinated by the brilliant white light produced by burning magnesium that he began an exploration of photochemistry. He developed his now-famous Bunsen burner

in order to produce a non-luminous flame that would not interfere with the spectra he and Kirchoff were studying. Through their work, Kirchhoff discovered that the characteristic bands of dark in the spectra produced by Fraunhofer's method could be matched to bands of colour produced by inverting the process.

If he illuminated the flame from behind with a bright light, Kirchhoff found, instead of dark bands in a bright spectrum, that there were bright bands of colour in darkness. In other words, the emission spectrum of a substance is the exact inverse of its absorption spectrum.

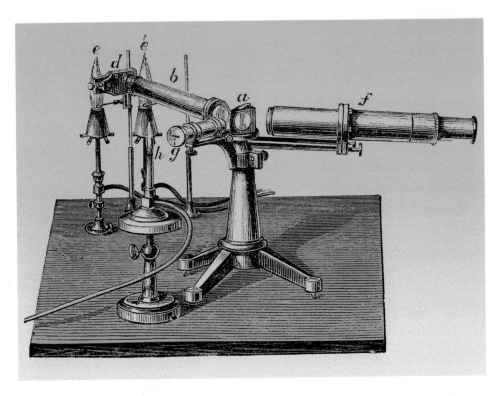

French physicist Léon Foucault had already seen emission spectra in 1849. Examining the spectrum from a voltaic arc between two carbon poles, he found a bright, double yellow line which matched the wavelength of the dark line that Fraunhofer had labelled the 'D' line in the spectrum of sunlight. After further tests, Foucault came to the conclusion that the arc would absorb light at the same frequency at which it emits light.

Bunsen and Kirchhoff set about logging as many spectra as possible, working late into the night in an enthusiastic frenzy of activity, throwing every substance they could find into the flame of their spectroscope.

In a landmark paper, 'Chemical analysis by spectral observations', published in

Above: This spectroscope uses Bunsen's specially designed Bunsen burner (h, e) to produce a non-luminous flame. The flame is viewed through a prism (a), looking through the tube (f).

1860, they recorded the spectra of lithium, sodium, potassium, calcium, strontium and barium salts. They also discovered new elements. By exposing Durkheim mineral water to the flame, they found caesium, and from the mineral lepidolite they discovered rubidium. Spectroscopy would become a valuable tool in the search for new elements.

Starlight, star bright

In their 1860 paper, Bunsen and Kirchhoff suggested that spectroscopy could be used to investigate the composition of the Sun:

improved four-prism spectroscope that allowed him to view the solar spectrum and the spectrum of the elements he was looking for side by side. He found evidence of iron, calcium, magnesium, sodium, nickel and aluminium in the outer layers of the Sun. He thought cobalt, barium, copper and zinc were probably present. The best evidence was for the presence of iron, for which he had 60 matching lines.

'It is evident that the same mode of analysis must be applicable to the atmospheres of the Sun and the brighter fixed stars . . . the solar spectrum, with its dark lines, is nothing else than the reverse of the spectrum which the Sun's atmosphere alone would produce. Hence, in order to effect the chemical analysis of the solar atmosphere, all that we require is to discover those substances which, when brought into the flame, produce lines coinciding with the dark ones in the solar spectrum.'

Kirchhoff set about examining the solar spectrum in great detail, building an

Lepidolite is a lilac-grey or pink mineral rich in lithium and other metals. Bunsen and Kirchhoff discovered rubidium through the analysis of lepidolite.

'At the moment, I am engaged in a research with Kirchhoff which gives us sleepless nights. Kirchhoff has made a beautiful and quite unexpected discovery: he has found out the cause of the dark lines in the solar spectrum, and has been able both to strengthen these lines artificially in the solar spectrum, and to cause their appearance in a flame spectrum, their positions being identical with those of the Fraunhofer lines. Thus the way is given by which the material composition of the Sun and fixed stars can be ascertained with the same degree of certainty as we can determine by means of our reagents the presence of sulphuric acid, chlorine, etc.'

Robert Bunsen, 1859

Staring at the Sun

A few years later, in 1868, an eclipse would give scientists the chance to test spectral analysis on the outer parts of the Sun. The chromosphere and corona are regions of the Sun's atmosphere that are visible only during an eclipse when light from the main body of the Sun is blocked.

This opportunity was seized by French astronomer Pierre Janssen, who went to Guntoor, India, to observe a total eclipse of the Sun. Like most observers, he concluded that hydrogen was certainly present. But among the bright lines in the spectrum that Janssen observed was a baffling yellow line that didn't match any known spectral pattern. Everyone had assumed it was sodium, but it was not identical to the sodium lines. Janssen set about developing an instrument that would enable him to examine the Sun's atmosphere, even when there was no eclipse, so that he could continue to study the problem.

Janssen was unaware that an English scientist, Joseph Lockyer, had embarked on the same task and managed to see solar prominences in October 1868. (Solar prominences are large, bright features that extend outwards from the surface of the Sun.) Lockyer also saw the rogue yellow line in the Sun's spectrum, and came to the conclusion that it represented a new element

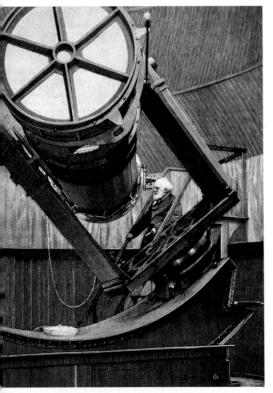

The astronomer Pierre Janssen in old age.

> **OUT OF THIS WORLD**
>
> While Janssen and Lockyer focused on the Sun, the husband-and-wife team William and Margaret Huggins investigated the composition of other stars with a spectroscope attached to a telescope. The Huggins couple, assisted by William Miller, trapped the spectrum of the star Aldebaran, finding 70 line positions. In 1864, they reported that the star contained sodium, magnesium, hydrogen, calcium, iron, bismuth, thalium, antimony and mercury. (They were wrong about the last four elements, having misread the lines.) They went on to study nebulae, coming to the conclusion that their very different composition suggested they were vast cloudy objects which would never resolve into stars. They published their *Atlas of Representative Stellar Spectra* in 1899.
>
> Their work demonstrated an important point: the stars, as well as our own Sun, are made of the same elements as Earth. Chemistry was shown to be universal.

not known on Earth. He named it 'helium', after *helios* the Greek word for 'sun'. Other chemists were not as excited or convinced as Lockyer and Janssen. The announcement that there was a new element, and that it could be found only on the Sun, was met with derision. It took nearly 30 years for the discovery of helium to be recognized, and then only after the discovery of a different noble gas – argon.

The missing air – and more

When Henry Cavendish listed the components of air in the late 18th century, he realized there was something missing but couldn't identify it. The missing component was argon, which makes up nearly 1 per cent of Earth's atmosphere. Eventually argon was discovered by William Ramsay, working in Scotland, and Lord Rayleigh (John Strutt) in England. The two men started off working independently, but were later in communication and are jointly credited with identifying the element in 1894.

Below: A prominence (top right) emerging from the surface of the Sun, captured by NASA's STEREO spacecraft in 2008. The prominence is a cloud of ionized helium. Cool in relation to the surface of the Sun, the cloud flares out then breaks apart and dissipates into space over a period of hours.

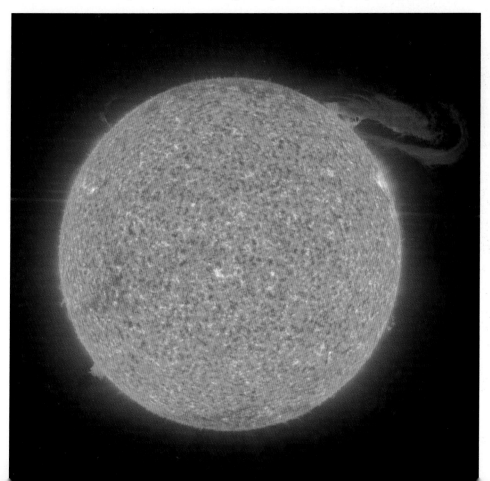

Lord Rayleigh found that if he extracted nitrogen from the air, it had a higher density than the nitrogen he made by decomposing ammonia. The difference was only small (1.257 g/litre for nitrogen from air against 1.251 g/litre for nitrogen from ammonia) but it merited investigation.

Ramsay suggested that Rayleigh look for a lighter gas polluting the nitrogen from ammonia.Meanwhile Ramsay looked for a heavier gas mixed in with the nitrogen from air. One or the other must provide the mysterious gas that was mixing with the nitrogen.

The oxygen tanks used by deep-sea divers contain a mixture of helium and oxygen, called heliox. It is also used medicinally, with patients who have breathing difficulties.

Ramsay removed all the nitrogen from his air-derived sample by passing it repeatedly over heated magnesium (with which nitrogen will react to form magnesium nitride). He was left with an un-reacting 1 per cent. It was denser than nitrogen, at 1.784 g dm^{-3}. This was just the right value to give the heavier density for nitrogen from air. Spectroscopy revealed red lines, and

Ramsay confirmed that the gas was a previously unknown element. He and Rayleigh proposed the name argon, which comes from the Greek *argos*, for 'lazy' or 'idle', because the gas is completely unreactive.

Grounded

One year later, Ramsay identified helium on Earth. He was working with uranium and found an inert gas that at first appeared to be argon. He sent a sample to Lockyer for spectroscopic analysis. Lockyer was delighted to find that it was, in fact, helium – a discovery that led to the proper recognition of helium as the first element discovered outside Earth. It remains the only one so far.

This presence of helium in uranium ore was a clue to how some still unknown elements might one day be tracked down. But it was not immediately useful, as in 1895 neither radioactivity nor the nature of the atomic nucleus was known. Likewise, no one had any idea why or how the elements could produce distinctive spectra. Chemists gratefully exploited the phenomenon without understanding it.

Ramsay's elements

One odd gas may be an anomaly, but two begins to look like a pattern. Ramsay suggested that argon and helium were representatives of a new group in the Periodic Table. The existence of elements with no immediately obvious place in the table initially distressed Mendeléev. He even argued that argon was not an element at all but molecular nitrogen; three atoms of nitrogen bound together. However, when it became clear that there was space

---

HELIUM: HERE TODAY, GONE TOMORROW

Helium (like hydrogen) has such a low mass that it readily escapes Earth's gravity and drifts off into space. It escapes so quickly that none of Earth's original supply of helium survives. There would be no helium at all on this planet if the supply were not constantly being replenished by activity deep within the Earth, where radioactive elements undergo decay by emitting helium nuclei (alpha particles – see page 144).

Alpha particles are helium nuclei: two protons fused with two neutrons. Having a positive charge (+2), they readily attract electrons, and become a helium atom. The helium eventually makes its way to the planetary surface; it's a small atom and can fit through the tiniest of gaps. The helium that Ramsay found in uranium-bearing ore had been generated by radioactive decay. Helium for industrial use is collected from underground natural gas deposits, principally in the USA (Texas, Oklahoma and Kansas).

for a whole group of elements between the halogens and the alkali metals, the noble gases became another factor that reinforced the accuracy of Mendeléev's design and he embraced them as Group 0. Since they fitted into the table, and as gaps in the consecutive atomic numbers were filled, Mendeléev's Periodic Table began to look more robust.

Now certain of their existence, Ramsay set about finding the rest of the noble gases. Alongside English chemist Morris W. Travers, he worked with air so cold that it had condensed to form a liquid. (Air had first been liquefied in 1883 by the Polish chemists Zygmunt Wróblewski and

Below: The priceless Waldseemüller map (Universalis Cosmographia) is the first one to mention America. Made in 1507, it is preserved in an inert atmosphere of argon in the Library of Congress, Washington, USA.

Karol Olszewski.) Their first discovery was krypton in 1898. This gas makes up only 0.0001 per cent of the Earth's atmosphere. Other components of the atmosphere are more volatile than krypton, so it remains a liquid when those with a lower boiling point have evaporated. (The boiling point of krypton is -153°C, while those of nitrogen and oxygen are -196°C and -183°C respectively.) Following closely on the heels of krypton, they found neon (boiling point -246°C), also through their work with liquid air. Neon is the fourth most common element in the universe, but on Earth it comprises only 0.0018 per cent of the atmosphere. Xenon, a third element found in the productive summer of 1898, also emerged. With a boiling point of -108°C, it is the least volatile of the components of air and accounts for only 0.0000087 per cent of the atmosphere.

Above: William Ramsay lecturing about noble gases.

a scant atmosphere of them could seem dense. He published his findings on two or more aetherial, pre-hydrogen gases filling interstellar space in a booklet entitled *An Attempt Towards a Chemical Conception of the Aether* (1904).

A tangle of weight and number

Mendeléev's idea of putting the elements in order of atomic weight but varying the order where it didn't quite work suggested two possibilities. One was that the atomic weights were sometimes wrong. The other was that atomic weight wasn't the true organizing principle after all.

Mendeléev plumped for the first theory. For most of the 19th century, Berzelius's table of atomic weights from 1828 formed the basis of calculations, and methods of measuring atomic weight were somewhat unreliable. Mendeléev urged chemists to check their measurements again and again. Often they discovered they had got it wrong and came back with better values. But a few pairs of elements remained stubbornly in the wrong order, with the lighter element coming after the heavier element in Mendeléev's table.

Mendeléev's proto-helium

In 1902, Mendeléev proposed an element lighter than hydrogen that would appear at the top of the new Group 0 gases and would possibly explain the newly discovered phenomenon of radiation (see page 160). He believed there were two possible elements lighter than hydrogen and suggested they might account for the *aether*, that elusive substance which has intermittently been deemed to occupy otherwise empty spaces since at least the time of Aristotle. Mendeléev considered that these super-light, almost massless gases permeate all matter but rarely interact with any of it. He believed that their particles moved extremely fast, so that even

Wet chemistry and Morley's oxygen

A much-needed breakthrough in measuring atomic weight came right at the end of the century. Edward Morley (1838-1923) was the most meticulous, multi-talented and skilful experimental chemist of his generation. When he turned to determining the atomic weight of oxygen, he did it with characteristic thoroughness and precision.

Morley used three different methods, which allowed him to cross-check his results. Like his predecessors, he focused on the reaction of oxygen with hydrogen to produce water. Unlike them, however, he weighed both of the reactants and the product (rather than weighing two of the three and depending on arithmetic to give the third value). Morley had also developed methods of producing purer hydrogen and oxygen. His value for the atomic weight of oxygen, 15.879, is remarkably close to the modern value for the oxygen-hydrogen atomic weight ratio, 15.8729:1.

Just as important as the accuracy of his measurement were the implications of the value. The atomic weight of oxygen was not a whole-number multiple of the atomic weight of hydrogen. This finally refuted Prout's hypothesis of 1815 that atoms of heavier elements are simply composed of hydrogen atoms (protyle) which have in some way been combined, condensed or compressed.

Better atomic weights

A reliable method of determining atomic weights was finally found in 1912 by British physicist J.J. Thomson (1856–1940) whose work led to the invention of the mass spectrometer. The method involved

sending ionized gas along a tube, where it was subject to both a magnetic and an electrical field. The electrical field changed the speed of the ions and the magnetic field changed their direction, bending their path. The ions were collected at the other end of the tube in 'Faraday cups', devices invented by Michael Faraday that were designed to catch charged particles in a vacuum and generate a current in wires attached to the cups. From the size of the current and the location of the ions' impact with the cups, it was possible to work out how far their path had been deflected and how much

The recognition that there is another number which is important in defining atoms emerged naturally from Mendeléev's work. He could simply number the elements sequentially once he had settled on their order. At first, though, atomic number was considered a rather arbitrary thing, with no identifiable root in the actual physics of the atom. It was just something that had a rough correlation with atomic weight and which worked conveniently to put the elements in the right order. The very real and non-arbitrary nature of atomic number could not be recognized until the nature of the atom was explored more thoroughly. The final piece in the puzzle was put in place by a young English physicist, Henry Moseley, in 1913 – the year after Thomson had perfected the method for discerning atomic weight.

their speed had been increased. Then, using Newton's second law, which states F = ma (force = mass × acceleration), it's possible to work out the mass of the particles.

But for all the precision Morley brought to the discussion of comparative atomic weights, the point really wasn't atomic weight at all – it was atomic number.

Seeing the light

The second of the two possibilities – that atomic weight wasn't the organizing principle after all – turned out to be the correct interpretation.

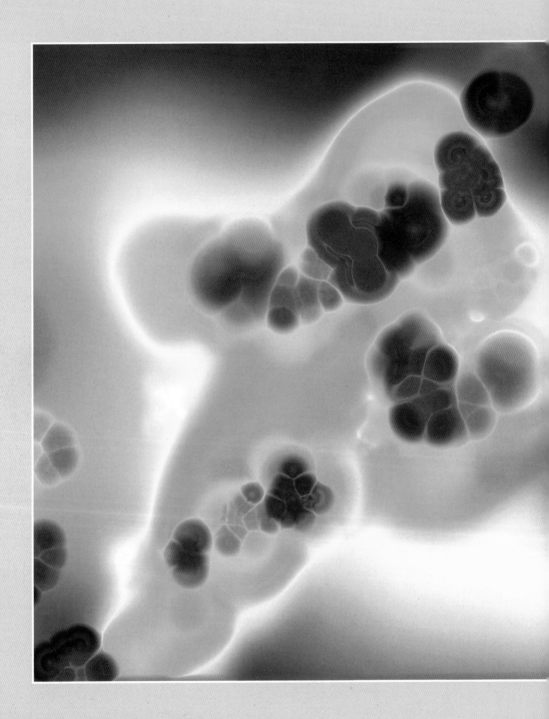

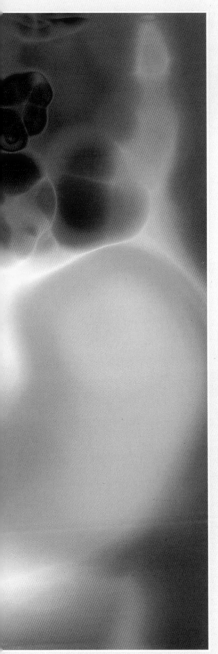

CHAPTER 7

ATOMS UNLOCKED

'We have here a proof that there is in the atom a fundamental quantity, which increases by regular steps as we pass from one element to the next. This quantity can only be the charge on the central positive nucleus, of the existence of which we already have definite proof.'

Henry Moseley, 1913

By the end of the 19th century most chemists believed that atoms existed. Most also accepted Lavoisier's model of an element as a substance with its own unique design of atom, and believed that atoms somehow hooked up to form molecules. What remained obscure was the nature of atoms themselves and the way they work, either together or alone. The answer to this puzzle would unlock the Periodic Table.

A transmission electron microscope can reveal structures at the level of individual atoms.

From rays to particles

Dalton described an atom as a 'solid, massy, hard, impenetrable, movable particle'. He was wrong with 'solid' and 'hard', as the vast majority of an atom is empty space (as Boščović foresaw). He was also wrong in saying that atoms were indivisible, not just on the cataclysmic scale of atom-splitting nuclear fission but on the mundane level of their comprising subatomic particles.

Dalton's conviction made it hard to see how atoms might fix together to form molecules. Berzelius tackled the problem, coming to the conclusion that some form of electric force held them together. He would turn out to be right, in a sense, but the route to the truth was circuitous. It began with the discovery of waves of electromagnetic energy in the middle of the 19th century.

Rays unravelled

Following the work of James Clerk Maxwell on electromagnetism, rays became a popular subject of intellectual enquiry in the second half of the 19th century. In 1865, Maxwell had demonstrated that electric and magnetic fields travel through space as waves, that light is a part of the same phenomenon, and that they all travel at the same speed. At first glance this doesn't look as though it has much to do with the elements, at least not beyond the use of light for spectroscopy. On the whole, physicists were more concerned with 'rays' than were chemists. But, in 1895, the discovery of X-rays by German physicist Wilhelm Röntgen brought physics and chemistry closer together.

Röntgen was exploring the path of what were known as 'electrical rays' from an induction coil through a partially evacuated glass tube. He found that the rays he was examining caused fluorescent material to glow. Further investigation revealed that they could penetrate some forms of material, but not others. And by using a photographic plate, Röntgen could produce an image

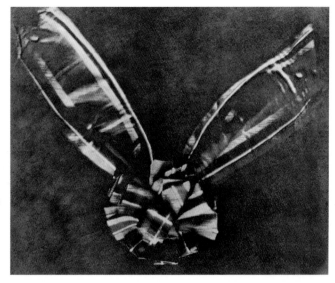

Following his work on light as electromagnetic radiation, James Clerk Maxwell produced the first colour photograph. His image of a tartan ribbon was produced by combining photos taken using red, green and blue filters.

The X-ray image of his wife's hand, made by Wilhelm Röntgen in 1895. Her wedding ring is clearly visible.

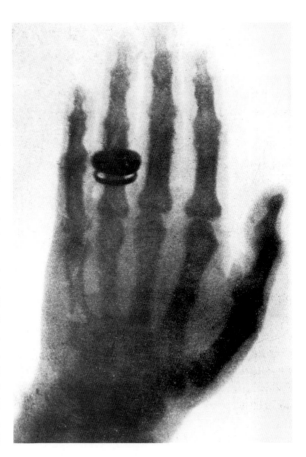

from the shadow cast by objects and materials impervious to the rays. He saw the bones of his own hand, and made the first X-ray image which was a picture of his wife's hand with her wedding ring clearly visible. Within a year, X-rays were being used in hospitals to reveal fractures and blockages.

The following year, French physicist Henri Becquerel tested his theory that phosphorescent materials such as uranium salts might emit a type of radiation akin to Röntgen's X-rays. Becquerel soon found that uranium does emit rays of some type, but they were not X-rays, as they could be deflected by a magnetic field. Exploring more widely, he found

URANIUM: NOT WHAT IT SEEMED

German chemist Martin Klaproth (1743–1817) discovered uranium in 1789. Or, rather, he believed he had. Klaproth was investigating the mineral pitchblende, which was thought to be an ore of zinc and iron, but is in fact mostly uraninite, an oxide of uranium (UO_2, with some U_3O_8). After dissolving the mineral in nitric acid and neutralizing it with potash, he found a yellow precipitate that dissolved if he added more potash. Klaproth concluded that pitchblende contained a new element; he named it uranium after the planet Uranus, which had been discovered in 1781 by William Herschel. On heating an oil-based paste of his precipitate in a charcoal crucible, Klaproth obtained a black powder with a metallic lustre, which he and subsequent chemists accepted as uranium. But uranium is a dull silver metal and Klaproth had produced only an oxide of it. There was a good reason that Klaproth and later chemists did not recognize the original product as an oxide: UO_2 can't be reduced using hydrogen or carbon.

there were three types of radiation; one type could be deflected in either direction and the third type could not be deflected at all, indicating that there were electrically positive, negative and neutral forms of rays/radiation (see also page 144).

Towards the electron

Electricity lay at the heart of Maxwell's work. Although electricity had been propagated chemically by Volta in 1800 (see page 92), no one really understood its nature. In the mid-18th century Benjamin Franklin had explained it in terms of positive and negative charges and scientists were putting it to good use by the middle of the 19th century. But the mediator of electricity, the electron, still lurked undetected.

Physicists and chemists investigated electricity in several ways. One of the most fruitful was to discharge it inside a glass flask containing a small amount of gas at low pressure. In 1869, German physicist Johann Hittorf found that a glow emitted from the negatively charged terminal (the cathode) increased as he decreased the pressure of the gas in the flask. In 1876, another German physicist, Eugen Goldstein, called the rays that produced the glow 'cathode rays'. English physicist and chemist William Crookes investigated them further, creating the 'Crookes tube'. This later became the cathode ray tube (CRT), used in television sets before LCD screens took over.

Crookes found that cathode rays went from the negative cathode to the positive terminal (or anode), and that the glowing beam could be deflected by a magnetic field. He called the substance of the rays 'radiant matter' and proposed that it represented a

Above: William Crookes, holding one of the gas-filled tubes that bore his name, the Crookes tube.

fourth state of matter (besides solid, liquid and gas) consisting of negatively-charged molecules which were forced out of the cathode at high speed. Arthur Schuster, a German-born physicist working in Britain, demonstrated that applying electrical potential across the beam deflected the rays towards the positive terminal. This indicated that the rays were negatively charged, as Crookes had proposed.

But still no one knew what the 'rays' were. Some people thought they were waves and others believed them to be charged atoms or molecules. In 1896, J.J. Thomson solved the puzzle. His experiments established that the rays consisted of a stream of 'corpuscles' as he termed them. He estimated the mass and charge of the corpuscles, finding them, astonishingly, to have less than one thousandth of the mass of a single hydrogen atom. He soon showed that their charge-to-mass ratio was the same no matter what material he used for the cathode, and that 'corpuscles' produced in different ways appeared to be identical.

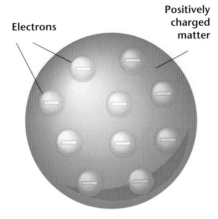

Electrons **Positively charged matter**

Left: Thomson's plum-pudding model of the atom had a sphere of positive charge with dots of negative charge (electrons) scattered throughout.

Below: The English plum pudding is a sphere of suet dough, studded throughout with currants and other dried fruit. In Thomson's atom, the part of the suet pudding was taken by a diffusive globule of positively charged matter and the part of the currants by electrons, whizzing around at high speed.

Time for a redesign

Thomson concluded that the corpuscles emanated from the diffuse gas in the tube – there was no other source for them. This meant he could break small bits off atoms. He set about redesigning the atom to take account of this discovery.

What he came up with has been called the 'plum pudding' model of the atom, after a traditional English pudding; his model would replace Dalton's theory of indivisible solid atoms. Thomson concluded that the main body of the atom must necessarily have a positive charge, since something had to balance the charge of the corpuscles to produce neutral atoms. The main body also had to account for the vast majority of the mass of the atom.

More rays

Of the three types of radiation identified, one would turn out to be Thomson's new corpuscles in the form of high-speed electrons.

In 1898, Ernest Rutherford (1871–1937), a New Zealand-born chemist, named the types of rays that could be deflected as alpha and beta radiation. He soon found that they are beams of particles rather than 'rays' of energy. Alpha particles are positively charged; beta particles (later identified as electrons) are negatively charged. The third type, which he named gamma rays, were actual rays. He announced the existence of 'alpha and beta rays' in uranium radiation, and listed some of their properties.

After moving to Manchester in England, Rutherford explored alpha particles further. In 1908–13, in a series of world-changing experiments, he directed Hans Geiger and Ernest Marsden in firing streams of positively charged alpha particles at very thin gold sheeting and then mapping where they came out the other side. Rutherford expected most of the particles would go straight through, though a few might be slightly deflected. But what happened was mind-blowing. Some of the alpha particles were deflected at a huge angle and a few even bounced straight back. Rutherford described it as being 'as if you fired a 15-inch shell at a piece of tissue paper and it came back and hit you.'

From pudding to planets

There was no way that the plum-pudding model of the atom could explain this finding. The model, therefore, must be wrong. If the atom was largely a diffuse cloud of positive charge, there would not be sufficient repellent force to deflect the alpha particles to such a degree. There must therefore be a high concentration of positive charge in a small space. The atomic nucleus was born, and with it a new model of the atom.

Rutherford worked out that the nucleus occupies only about one ten-thousandth of the diameter of an atom, so electrons roam a long way into the space around

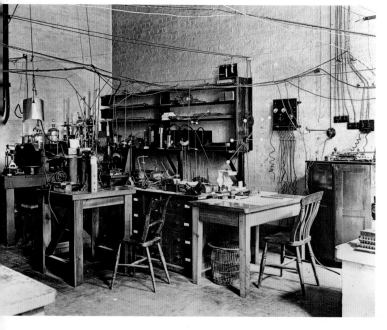

Left: Ernest Rutherford's research room in the physics laboratory of the University of Cambridge.

it. Most of the atom was, then, empty space with a small positively charged nucleus and particles of negative charge around it. This explained Geiger and Marsden's results: most alpha particles went straight through the gold sheeting without encountering any atomic nuclei, because they occupy such a small space. But a few alpha particles encountered the positively charged nucleus and were violently repelled, a reaction that changed their course dramatically. Heavier atoms deflected the alpha particles further, as their nuclei have a higher concentration of positive charge. Rutherford publicized this surprising experimental result in 1911.

Above: In Bohr's planetary model, electrons move in fixed orbits.

Two years later, Danish physicist Niels Bohr (1885-1962) proposed a new model of the atom, which is familiar today. In Rutherford's experiment, the electrons wandered a good distance from the compact nucleus, but he didn't constrain their movement. Bohr, however, gave the electrons specific orbits from which they were not allowed to stray. In his theory, called the planetary model of the atom, the bulk of the atom's mass was concentrated in the positive nucleus and surrounded by a different number of electrons for each of the elements, arranged in fixed orbits around it. The orbits could extend a long way into space, so the atom could occupy a very large volume relative to the diameter of the nucleus. The model was later refined so that the orbitals were defined as energy levels rather than specific physical locations. With the emergence of quantum theory, it became clear that the physical location of an electron can never be precisely identified. The orbitals are really realms of probability: the electrons are most likely to be found in those areas.

Atoms are real

The first evidence for the existence of atoms is generally dated to 1827 when an English botanist, Robert Brown, was examining pollen under a microscope. Brown saw tiny particles that appeared to move randomly; the paths of some of the particles were temporarily distorted as though they were being pushed from one side. At first he supposed that the particles were animate and moving of their own volition. To test this, he examined pollen taken from dead plants – some over 100 years old. Still the particles moved. He went on to test many kinds of substance that had no living content, including coal, glass and metal. In all of them, the particles moved. Brown called his tiny moving particles 'molecules' and measured some as small as one-thirty-thousandth of an inch.

Initially, no one could explain Brownian motion as it became known. But in the

Robert Brown was examining pollen grains of the plant Clarkia pulchella *when he observed the movements later named Brownian motion and taken as evidence of the existence of atoms.*

middle of the 19th century, emerging kinetic theory offered a model.

A fluid is made up of invisibly small molecules which are in constant motion. When a small, solid particle is suspended in liquid, moving molecules continuously collide with it, jostling it around. As the path taken by the molecules is random, sometimes there will be enough force pushing on one side to produce visible movement. At other times, bumps from different sides cancel one another out. As Brownian motion increases with increasing temperature, this kinetic theory seemed to offer a satisfactory explanation – at least for those willing to entertain the idea of atoms and molecules.

The issue was finally settled in 1908 by Austrian physicist Albert Einstein. He produced mathematical models for the movement of particles in liquid at different temperatures which were tested experimentally by French physicist Jean Perrin. Einstein's predictions were confirmed by Perrins' results and provided the first experimental evidence of the existence of molecules. Finally, after 2,500 years, the question about the atomic or continuous nature of matter was resolved.

Charges, numbers and atoms

So far, none of this related in any way to periodicity. That would change when physicist Henry Moseley (see page 149) became convinced that atomic number must somehow relate to the physical state of the nucleus.

In 1911, Dutch amateur physicist Antonius van den Broek suggested that atomic number related to a real feature of atoms, the charge in the nucleus: 'If all elements be arranged in order of increasing atomic weights, the number of each element in that series must be equal to its intra-atomic charge.' Van den Broek had no evidence to back up his suggestion; Moseley set about proving it correct.

Moseley worked with Rutherford in Manchester, and started with the latter's model of the atomic nucleus of negative electrons orbiting around a dense, positively charged nucleus. Rutherford had stated that the charge on the nucleus was large, at about half the atomic weight of the atom. He suspected that the nucleus might be composed of helium nuclei, which have a mass of 2, and that it would have a charge of 1. But his idea didn't work rigorously throughout the Periodic Table. For example, gold was in position 79 but Rutherford found it to have a charge of about 98. Because of the discrepancies, he made no public claim linking the charge on the nucleus and atomic number.

Left: A map of particles moving under Brownian motion shows that the path is entirely random.

Moseley set about his task using X-ray spectroscopy (see page 150). The English physicist Charles Barkla had demonstrated that the elements emit characteristic X-rays, which he called K and L rays. This meant that the X-rays related in some way to the nucleus, as they varied from one element to another. Moseley fired a beam of electrons at a series of elements and measured the wavelength of the K X-rays each emitted. He began with ten elements, between calcium and zinc, but excluding scandium because he couldn't get a sample of it. Sure enough, he found a linear relationship that matched their relative atomic numbers. Plotting atomic number against the square root of the frequency of K-rays showed a clear correlation with the X-ray frequency, increasing in steps of a fixed size as atomic number increased. In 1914, Moseley extended his study to test most of the elements between aluminium and gold. The results were consistent with his earlier findings.

The nature of the nucleus was still not fully understood, so the reason for the regular steps in K-rays could not be explained. However, Moseley's discovery demonstrated conclusively that atomic number is a genuine feature of the atom, not just a convenience that helps chemists put the elements in order.

The next step was to work out what in the atom gave the atomic number: what did atomic number actually mean?

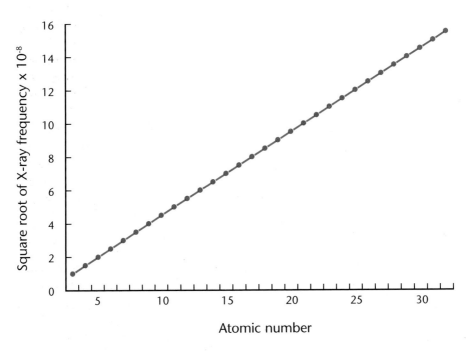

Moseley's graph of atomic number plotted against the square root of X-ray frequency demonstrated conclusively that atomic number is more than a convenience devised by chemists.

HENRY MOSELEY (1887–1915)

Henry Moseley's mother was the daughter of a prominent biologist. His father, an Oxford science professor, died while his son was still young. Moseley showed early promise, winning a scholarship to Eton. In 1906 he won prizes in physics and chemistry before going on to study at Trinity College, Oxford. After graduating, he worked with Ernest Rutherford at the University of Manchester, first teaching physics and then as a researcher.

In 1912, he demonstrated that radioactivity can be used to power a battery. His method consisted of shielding radium from the beta particles (electrons) it emits, so that the charge of the radium becomes positive. Moseley hoped to achieve a charge of one million volts, at which point the beta particles would be drawn back immediately, but he was unable to insulate his source sufficiently. Even so, he generated 150,000 volts, making the first radioactive battery, which he called a radium battery. Radioactive batteries are now used in situations where long battery life is crucial, such as for cardiac pacemakers and spacecraft.

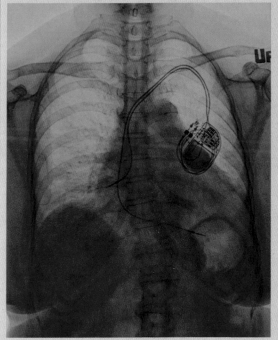

Moseley returned to Oxford where he put together the X-ray gun he needed to investigate atomic number. He completed this work within a year. This brilliant start to his career would be his greatest achievement. At the outbreak of war, in August 1914, Moseley enlisted with the British army and was sent to Turkey as a technical officer in communications. In August 1915, at the Battle of Gallipoli, he was shot in the head by a sniper. At the age of just 27, one of the most promising scientists of his generation was dead.

Left: A pacemaker, powered by a radioactive battery, shows up on an X-ray of a patient's chest.

WEIGHT V. NUMBER

Atomic weight (or relative atomic mass) is the average mass of an atom in a particular sample of an element, given as multiples of 1/12th the mass of a carbon-12 atom. The definition refers to 'a particular sample' because many elements exist as isotopes — that is, in forms with different numbers of neutrons, so with atoms of different mass.

Atomic number refers to the number of protons in the nucleus of an atom. The number of electrons is always the same as the number of protons, so the atomic number indicates the valence (combining power) of an element. Atomic number is always a whole number; no atom can have part of a proton.

Gaps and phantoms

Moseley's graphs have gaps where either an element was difficult to provide in a form for use with the X-ray spectroscope, or where an element had not yet been discovered. He gave no figure for any of the noble gases, which are extremely unreactive. One result was that – like Mendeléev – Moseley could predict the existence of several elements not yet discovered. He found gaps in his graph that corresponded to the elements with atomic numbers 43, 61, 72 and 75. These elements are now known as technetium (discovered in 1937), promethium (probably produced in 1942, but separated in 1945), hafnium (1922) and rhenium (found in 1908, but misidentified as element 43; properly identified when rediscovered in 1925). Two

Below: Examining the X-rays emitted by a sample reveals its composition. This is the basis of X-ray spectroscopy, a widely used technique in chemical analysis. It is even used by Mars rovers to examine rocks and dust on the planet.

of these, elements 43 and 75, correspond to elements Mendeléev also predicted.

Moseley also set about solving other puzzles about the Periodic Table. For some time, scientists had pondered the possibility of one or even two elements lying between hydrogen and helium. Helium has four times the atomic weight of hydrogen, so there appeared to be space for two further elements, with an atomic mass of 2 and 3. The two phantom elements had even been named: nebulium and coronium. But this line of speculation was swiftly curtailed by Moseley's discovery. Atomic number, not atomic weight, is the determining factor. As the atomic number of helium is 2 and lithium is 3, there is no space for more elements.

CORONIUM AND NEBULIUM

In 1864, William Huggins identified an unexplained green line in the spectrum of the Cat's Eye nebula.

In 1868, when helium was found in the spectrum of the Sun and accepted as an element, the prospects of Huggins' green line looked good and the name nebulium was proposed for this potential new element.

In 1869, American astronomers Charles Augustus Young and William Harkness independently found a bright green line in the spectrum of the Sun's

The green light produced by highly ionized iron is so pronounced it is sometimes visible to the naked eye during a solar eclipse

corona during a total eclipse. They assumed the line had been produced by a new element, and they named it 'coronium'. Sixty years later, Swedish astronomer Bengt Edlén concluded that the line was produced by iron under extreme pressure, rather than than by a new element.

In 1927, American physicist and astronomer Ira Bowen realized that nebulium is doubly-ionized oxygen. This can't exist on Earth, but it is common in nebulae.

Electrons and protons

As we have seen, Moseley discovered the connection between the charge on the nucleus (so the charge provided by protons) and the atomic number of an element. As the number of protons corresponds exactly with the number of electrons, this provides the link between the properties of the elements observed and their position in the Periodic Table. The chemical and physical properties of the elements are produced by the configuration of electrons. Effectively, Moseley had solved the puzzle of the elements, but exactly how and why would only emerge fully with further work on the structure of the atom that was carried out after his untimely death.

It's hard to say exactly when the proton was discovered. Rutherford had postulated a nucleus with a positive charge that balances the negative charge of the electrons in 1911, but didn't specify that it contained individual particles which correspond one-to-one to the electrons. When he found in 1917 that he could liberate hydrogen nuclei by bombarding nitrogen with alpha particles (helium nuclei), he determined that the hydrogen nuclei must be components of nitrogen. The nitrogen, losing a hydrogen nucleus, was becoming oxygen-17, following the reaction: $^{14}N + \alpha \rightarrow {}^{17}O + p$. This was the first recorded nuclear reaction. Rutherford found that the positive charge of any atomic nucleus could always be counted in whole numbers of hydrogen nuclei. He suggested the name 'proton' in 1920, borrowing from Prout's hypothetical protyle.

There was one piece of the atomic puzzle still undiscovered: the neutron. Although Rutherford proposed that there must be an additional particle in the nucleus, with

The path of alpha particles fired into helium, showing one collision (left); and of alpha particles bombarding nitrogen to produce oxygen and a freed proton (right).

no charge and a mass number 1 (equal to a proton), this was not discovered until 1932.

A new bit of atom

Rutherford's work on disintegrating atoms returned repeatedly to the fact that atomic mass is always greater than atomic number. As atomic number is the number of protons in the nucleus, this implies there is something else present in the nucleus to make up the extra mass. James Chadwick, a young researcher and World War I veteran, was working with Rutherford at Cambridge. He kept the problem of the missing mass at the back of his mind.

Meanwhile, in France, Irène and Frédéric Joliot-Curie (Marie Curie's daughter and her husband) were using new methods for tracking radioactive particles. Inspired by their work, Chadwick repeated their experiments looking for evidence of a new nuclear particle. In 1932 he found it, the neutron, and showed that the proton has 99.9 per cent its mass. German physicist Werner Heisenberg demonstrated that this was not a proton–electron pair, but a new, independent type of subatomic particle.

Reactivity explained

Why do some elements form compounds readily and others not? Why do they combine in different proportions? As soon as Thomson had discovered the electron, it became possible to explore how atoms might join with others to form compounds

1	2	3	4	5	6	7	8
H•							•He•
Li•	•Be•	• B •	•C•	•N•	•O•	:F•	:Ne:
Na•	•Mg•	•Al•	•Si•	•P•	•S•	:Cl•	:Ar:
K•	•Ca•	•Ga•	•Ge•	•As•	•Se•	:Br•	:Kr:
Rb•	•Sr•	• In •	•Sn•	•Sb•	•Te•	: I •	:Xe:

Above: Gilbert Lewis's proposed arrangements of atoms in the outer shells of some of the elements of the Periodic Table.

and why they bond in the way they do. Chemists began to look for what makes the elements the way they are.

Cubes and corners

American chemist Gilbert Newton Lewis (1875–1946) devised a model of how atoms might combine using their electrons as some form of 'glue'. While explaining valency (the combining power of elements) to his students in 1902, Lewis used a diagram in which he depicted atoms as cubes, with electrons at their vertices. This gave each atom enough space to house eight electrons. There could be concentric cubes, so that an atom was like a series of nested boxes, but only the outermost layer was involved in forming bonds. Lewis conjectured that if an atom had no electron at one of its vertices, it would need to get another electron to occupy that space. Similarly, if an atom had two unoccupied vertices, it would be able to take on another two electrons.

Lewis published his ideas in 1916 in a landmark paper laying out the basis of covalent bonding: that atoms form bonds by sharing electrons. By this time, Bohr had published his model of the atom. Instead of electrons sitting on the corners of cube-shaped atoms, they were orbiting a central nucleus in their designated areas. This allowed for an electron to be shared quite easily, as it could orbit two nuclei at the same time, binding the atoms together. Thus valency relates to the number of electrons it is able to share, donate or receive.

Lewis made the distinction between covalent and ionic bonds in 1923, defining an acid as any substance that will receive one or more electrons. He defined a base

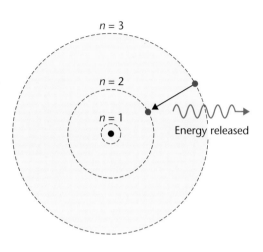

Above: In Bohr's model of the atom, an electron can only occupy set orbitals. It emits energy as it moves from a higher-energy orbital to a lower-energy orbital (one closer to the nucleus).

IONIC AND COVALENT BONDS

Chemists recognize two ways in which atoms form bonds to make molecules. In an ionic bond, one atom gives away one or more electrons and another atom accepts the electrons. Each atom aims to have a complete outer shell. In sodium chloride (table salt), for example, sodium has one electron in its outer shell, and chlorine has seven electrons in its outer shell. Sodium donates its electron to chlorine, which then has a complete outer shell. Both then have a satisfactory complete outer shell.

In a covalent bond, atoms share electrons between them. In the hydrogen molecule, for example, which consists of two hydrogen atoms, each shares its single electron with the other, with the result that both can benefit from a completed shell of two electrons.

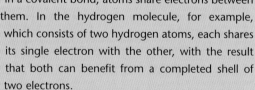

Left: A hydrogen molecule (H_2) has a covalent bond joining two hydrogen atoms.

as a substance which donates electrons. Both acids and bases aim to complete their outer shells by filling gaps or losing spare electrons.

Ordering electrons about

Bohr's model of the atom, with its electrons locked into their orbitals, fitted well with Lewis's ideas about electrons as the mediators of bonding. Together they fully explained the patterns of reactivity seen in the groups of elements as they were laid out in the Periodic Table. In 1923, Bohr's work on the application of quantum theory to atomic structure thoroughly nailed the issue and further explained the patterns of light gained with spectroscopy (see page 126).

Bohr's explanation relied on electrons occupying different orbitals within their shells, and with different energy profiles associated with those orbitals. An atom will fill its electron orbitals, starting with those of lowest energy (nearest to the nucleus) and progressing to those of higher energy. (This is the Aufbau principle, formulated by Bohr and Austrian-born physicist Wolfgang Pauli in the early 1920s.)

Yearning for completion

Atoms are stable (unreactive) when all their orbitals in occupied levels are filled. The most stable elements are the noble gases, which have their full complement of electrons in all their levels.

Bohr realized that the periodic nature of the properties of elements reflects the number of electrons in an atom's outermost shell. The alkali metals, for example, each have only one electron in their outer shell. Sodium has a configuration of 2.8.1 – the outer one is the electron it donates to chlorine to form sodium chloride. Lithium has an electron configuration of 2.1 and potassium of 2.8.8.1. Like sodium, they are highly reactive as they need to get rid of that spare outer electron.

The halogens, on the other hand, all have seven electrons in the outer shell and will only be stable if they can gain an electron to fill it. The halogens and alkali metals readily react together to form ionic bonds, fulfilling the need of both atoms to gain a complete outer shell, one by gaining an electron and the other by losing one.

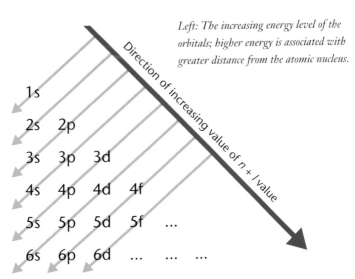

Left: The increasing energy level of the orbitals; higher energy is associated with greater distance from the atomic nucleus.

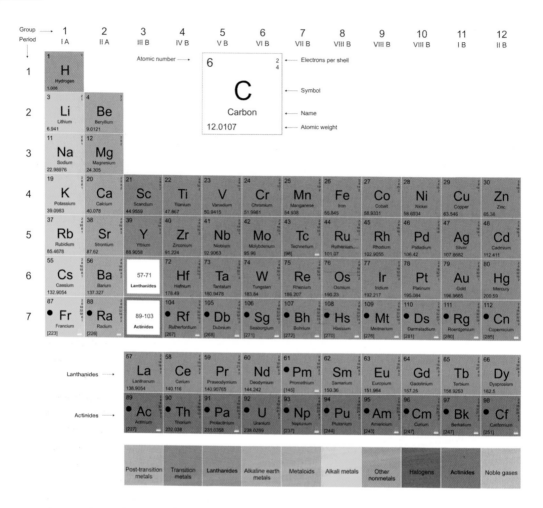

Atoms have layers – like onions

The energy levels which electrons can occupy are labelled from n=1 (closest to the nucleus) to n=7 (furthest from the nucleus, and occupied only in the elements of period 7). As we move down the Periodic Table, an extra energy level is added to the atoms at each row. The first row consists of hydrogen and helium only. These atoms have electrons only in level n=1. Hydrogen has one electron and helium has two. This level, uniquely, is complete with just two electrons. Hydrogen achieves this by combining with another hydrogen atom to form a diatomic molecule (H_2).

The levels form concentric spheres around the nucleus. Obviously, moving further from the nucleus, a spherical level encompasses a greater volume. While the first level has space for only two electrons, the next level, n=2, has space for eight. These are arranged in four pairs. Atoms don't fill their orbitals from the centre outwards within a level. Electrons repel one another

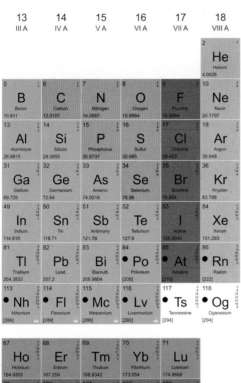

13	14	15	16	17	18
III A	IV A	V A	VI A	VII A	VIII A

Left: The Periodic Table as it stands now, from hydrogen to oganesson.

In Moseley's experiment, bombarding an element with a stream of X-rays produced K X-rays. The amount of energy given out (the wavelength of the X-rays) depends on how strongly the electrons are attracted to the nucleus. The more protons an atom has in its nucleus, the higher the positive charge. Therefore, the number of protons (the atomic number) determines the energy of the X-rays, and, as the atomic number increases, the wavelength of the X-rays produced increases correspondingly.

Back to mass and number

It might seem that if the atomic mass of an atom is the mass of the protons plus the mass of the neutrons, and these two have the same mass, the atomic weight must be exactly twice the atomic number. Yet, as Rutherford noted, this is often not the case, particularly with larger atoms. In fact, an atom's mass is not the sum of the mass of its parts. Some mass is lost in binding energy—the energy it takes to hold the nucleus together.

As Einstein showed in his famous equation, $E=mc^2$, energy and mass are ultimately interchangeable. Some of the mass of the components is converted into energy. This represents a loss of mass, so the atom has less mass than we would expect. The atomic weight figures shown on the Periodic Table are mostly more than twice the atomic number because they are an average figure based on the mix of isotopes of the element (see page 168).

as they all have negative charge, so they lie as far apart as possible. Consequently, each orbital first gets one electron, then when each orbital has a single electron, they gain a second in turn, starting with the orbital nearest the nucleus.

Moseley's work explained

Bohr's work made it possible to explain Moseley's discovery. When an electron falls from a higher energy level to a lower one, energy is released as electromagnetic waves.

CHAPTER 8

MADE TO MEASURE?

'Scientific work must not be considered from the point of view of the direct usefulness of it. It must be done for itself, for the beauty of science, and then there is always the chance that a scientific discovery may become like the radium, a benefit for mankind.'

Marie Curie, 1921

The aim of the alchemists was the transmutation of matter: making one substance from another, fundamentally changing its nature. The 20th century saw their dream fulfilled. But it has not brought limitless wealth. One application of modern 'alchemy' has the potential for almost unlimited destruction.

The vast explosions caused by nuclear weapons are created by the energy released by transforming elements.

New ways of looking

Throughout history, chemists have searched for new substances, and then new elements, in the world around them. For centuries their principal tools were fire and solvents. By the start of the 20th century most of the elements that could be found with these tools had been discovered. But chemists seeking new elements had two valuable new tools: spectroscopy and radioactivity.

Spectroscopy helped scientists to work out whether a substance was a new element. A spectral signature which couldn't be matched to a known element indicated that something new had been discovered. And radioactivity provided a means by which some elements came into being. A new question emerged from this: if elements occurred naturally by radioactive decay, could they be made artificially by the same means? The 20th century would see the first synthetic elements – those forced into existence by humans tampering with the very roots of matter.

X-rays, U-rays and a dark drawer in Paris

In 1896, French physicist Henri Becquerel discovered radioactivity while working on phosphorescent materials (materials that glow in the dark, emitting light they have previously stored). He was hoping to establish a link between phosphorescence and X-rays, but instead found that various uranium compounds could turn a

Below: One of Henri Becquerel's photographic plates, which led him to the discovery of radiation.

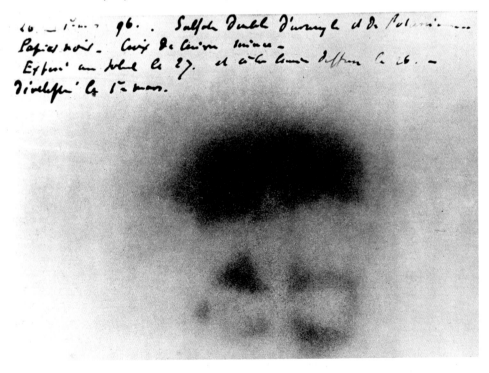

THE ELECTROMAGNETIC SPECTRUM

The electromagnetic spectrum is the full range of radiative energy. The light we can see (visible light, from red to violet) is part of this spectrum, as are the microwaves that heat our food and the radio waves that carry broadcasts and our wireless internet. The difference is simply in the wavelength – the distance between two peaks or two troughs in the waves of energy. X-rays and gamma radiation are also parts of the electromagnetic spectrum. Alpha and beta radiation are not, as they are moving particles rather than waves of energy.

Right: The electromagnetic spectrum, from the longest wavelength (radio waves) to the shortest wavelength (gamma rays).

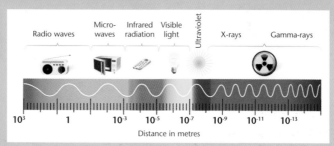

---

photographic plate black. The discovery, like so many, was entirely accidental.

Becquerel had been exposing uranium salts (potassium uranyl sulphate) to sunlight, then placing them next to photographic plates wrapped in black paper. On developing the plates, he found shadows on them created, he concluded, by some kind of 'rays' emitted by the salts. When the weather changed and became too dull for his experiments involving the absorption of sunlight, Becquerel placed the uranium salts and photographic plates in a drawer for later. After several days he took the plates out of the drawer and developed them, even though he had not carried out any experiments with light-exposed uranium salts. To his astonishment, the 'rays' had penetrated the paper and left their mark, even though the uranium salts had not had the chance to absorb light energy. He found that metallic uranium had the same effect.

Becquerel initially considered the rays to be something similar to X-rays, but further examination showed they were very different. He built an apparatus to deflect the emissions, using a magnetic field. X-rays can't be deflected by a magnetic field, but his rays could. Becquerel called them U-rays, and thought of them only in the context of uranium. Others referred to them as 'Becquerel rays' and were prepared to look further afield to find them.

Hidden elements emerge

It soon emerged that uranium was not the only element to emit Becquerel rays. Becquerel's student, Polish chemist Marie Skłodowska Curie (see page 163), discovered that thorium also produced U-rays. She and her husband Pierre worked with the uranium ore called pitchblende and found it to be more radioactive than uranium itself.

Pitchblende, a form of the mineral uraninite (an oxide of uranium), from which uranium is extracted.

paid off, and they discovered two new radioactive elements. They isolated the first, polonium, in 1898, naming it after Marie's homeland. It turned out to be one million times more radioactive than uranium. But the residue remained radioactive after the polonium had been removed, strongly suggesting there was yet another radioactive element present.

Pitchblende was expensive, as it was used to harvest uranium, and the Curies needed enormous quantities of it. Marie contacted a uranium factory in Austria, which sold her several tonnes of pitchblende from which the uranium had been extracted. The Curies set to work with the cauldron again, finally isolating the second element, which they called radium, in 1902. Radium turned out to be 2.5 million times more radioactive than uranium. Extracting polonium and radium was a considerable feat: a ton of uranium ore yields only 0.14 g of radium and just 100 micrograms, or 0.0001 g of polonium.

Even the waste material left after the uranium had been extracted was more radioactive than uranium. Concluding that pitchblende must contain at least one other radioactive element, they investigated further, melting down thousands of kilograms of pitchblende in an enormous cauldron and stirring it with an iron rod almost as tall as Marie herself. The effort

Following the pattern of phosphorescence, of the absorption and release of energy, Marie suggested that radioactive materials absorbed some kind of perennially present background radiation; they can then emit this, as phosphorescent materials absorb and emit light. It was the Curies who coined the term 'radioactivity'.

SAFETY LAST

Stirring a pot of heated, hugely radioactive material is not a healthy pursuit. Marie eventually died from aplastic anaemia, possibly brought on by exposure to radioactivity throughout her career. In the Curies' laboratory in Paris, a cavalier attitude towards potential risk was endemic. Concerns about safety were considered evidence that a person was not sufficiently dedicated to science and progress. Many of Marie's researchers fell sick. She and Pierre were already showing early signs of radiation sickness by 1903.

MARIE CURIE (1867–1934)

Marie Curie was born Maria Skłodowska in Poland to school teacher parents. Her education was interrupted when her mother died. Maria worked as a governess to help the family finances, reading and studying avidly in her free time. In 1891, she moved from Warsaw to Paris and enrolled at the Sorbonne to study maths and physics.

She met Pierre Curie and married him in 1895, adopting the French 'Marie' for her first name. While working at the School of Chemistry and Physics in Paris, the pair discovered that radiation was not affected by the chemical state of an element: a radioactive atom would emit radiation at the same rate and in the same way regardless of whether it was in a compound or in its native form. For their work in recognizing that the physics of radioactivity is separate from the chemistry of the elements, they shared a Nobel prize with Becquerel in 1903.

Their continued work with uranium led them to the laborious discovery of polonium and then radium. Marie won a second Nobel prize for the discoveries, this time in chemistry, in 1911. After Pierre's death in 1906, Marie continued their work, succeeding her husband as professor of chemistry at the Sorbonne.

When World War I broke out, Marie and her 17-year-old daughter Irène (later to become a radiochemist in her own right) took mobile X-ray wagons to the front line in France and worked diagnosing fractures and finding shrapnel in wounded soldiers. Marie later pioneered the use of radioactivity in medical treatments, particularly for cancer. She was the first person to win two Nobel prizes and remains the only woman ever to have done so.

Marie and Pierre Curie pictured with their daughter Irène, who also went on to win a Nobel prize for her work on radioactivity.

Radium horrors

The dangers of radioactivity weren't slow to emerge. In 1901, Becquerel sustained burns from radium he was carrying in his pocket. By 1904, Thomas Edison's X-ray assistant Clarence Dally had died of cancer, despite having had both arms amputated in an attempt to stop its spread. But ignorance and a conspiracy of silence prevented suspicions about radium's safety having any impact on the way it was handled.

Products laced with radium were sold as health-giving in the early 20th century. Items augmented with radium included milk, butter and chocolate, and cosmetics such as face cream, eye shadow and lipstick (which claimed they would make the user's beauty 'really shine'). Radium was even added to lingerie to enhance a couple's sex life. People popped radium pills for glowing health and put radium potions on their heads to restore greying hair to its former colour. In hospitals, surgeons stitched little capsules of radium into the surgical wounds of post-operative cancer patients.

It took a while for the popularity of radium to dwindle. Despite its risks, a

Advertisements promoted products from cosmetics to lingerie containing radium.

greedy industry keen to protect its profits covered up the element's worst effects. From 1917, the United States Radium Corporation (USRC) factory in Orange, New Jersey, employed women to paint the dials of clocks and watches with radium paint so that they glowed in the dark. The women would lick their radium brushes to

---

THE FIRST DEATH

USRC employee Mollie Maggia was only 24 years old when she died of the horrific effects of radium poisoning. First, she suffered toothache and had a tooth removed. But the pain moved to other teeth – she lost those, too. Her mouth was soon crowded with seeping, stinking ulcers. Before long, her entire jaw, mouth and part of her ear was a mass of pus and blood. Her jawbone broke apart in her dentist's hand. By this time, all her limbs ached unbearably and she could no longer walk. At last, the disorder ruptured the jugular vein in her neck and blood flooded her mouth. She was dead within a year of the first symptoms appearing. Other workers had begun to show the same signs. The cause of death was recorded as syphilis, an error that worked in the USRC's favour, as they could then deny responsibility.

achieve a fine point for detailed work. Even after doubts began to surface about the safety of radium, the Corporation continued to assure the women that their health was not at risk. At the same time other (male) workers in the same factories used lead aprons for protection when handling larger quantities of radium.

The women workers began to lose teeth and grew massive tumours on their faces; they suffered spontaneous bone fractures and fell sick in various other ways. The first death was in 1922. Everyone was puzzled until, in 1925, a medical doctor called Harrison Martland discovered that the human body treated radium in the same way that it treats calcium: radium was being laid down in the women's bones, replacing calcium and doing damage to other tissues.

Painting the faces of clocks doesn't look like a dangerous job, but it was deadly for the women working for the US Radium Corporation.

The factory owners dismissed the dangers, tried to quash all research, and continued to assure workers they were not in danger. The 'radium girls', as they became known, who fought for justice had an uphill struggle, and not just against those with a vested interest. During the Great Depression it was deeply unpopular to attack one of the few businesses that was still solvent and offering employment.

The girls finally won their case in 1939, when the radium companies were found guilty of gross negligence. Legal changes to protect other workers followed directly from the USRC case.

Transmutation at last

It took only a few years to solve the mystery of radioactivity. Just four years after Becquerel's discovery, in 1900, Rutherford and English chemist Frederick Soddy, working at McGill University in Montreal, Canada, recognized that radioactive decay involves the transmutation of one element into another. It was what the alchemists of old had sought to achieve. One element slowly changed to a completely different element, with the emission of 'Becquerel rays' a vital part of the process. They published their findings in 1902.

Rutherford observed that different radioactive elements decay at different rates, and coined the term 'half-life period' in 1907 to report the rate of decay: the average time it takes for half of a given sample of a radioactive material to decay into another element. (We now call it just 'half-life', but the plural is correctly half-lifes from 'half-life periods', rather than half-lives.) Rutherford also introduced the idea of judging the age of rocks by investigating how much radium had decayed to lead-206.

Here today, gone tomorrow

A natural consequence of radioactive decay is that the existence of some elements is temporary. In the case of an element such as uranium, 'temporary' is a relatively loose term, as uranium-238 has a half-life of 4.5 billion years. This is roughly the age of the solar system.

Other radioactive elements have much shorter half-lifes, which makes it remarkably easy to miss them. The gas

---

AS OLD AS THE HILLS

Argon is a product of the radioactive decay of potassium-40, which has a half-life of 1.25 billion years. This very long half-life makes the potassium-argon balance a useful way of dating geological and archaeological samples. When volcanic rock is hot and liquid, gas escapes from it. Once it cools, any gas produced by the radioactive decay of potassium is trapped. By measuring the proportion of argon in rock, scientists can judge its age (that is, the time since it solidified) to within about a million years.

ALPHA, BETA, GAMMA

Alpha radiation is the emission of an alpha particle from the nucleus of an atom. An alpha particle is two protons and two neutrons – essentially, the nucleus of a helium atom. It is the least penetrating form of radiation and can be stopped by a sheet of paper or a few centimetres of air.

Beta radiation is the emission of high-energy electrons (or positrons, particles like electrons but with a positive charge). It can travel several metres in air and penetrate skin, but can be stopped by a thick piece of plastic.

Gamma radiation is emitted as photons, acting as waves. As photons are not particles with mass or charge, they can travel hundreds of metres in air. They can be stopped only by a thick sheet of a dense material, such as lead.

radon was discovered in 1900 by German chemist Friedrich Dorn while he was investigating the decay chain of uranium. He named it 'radium emanation', but it was renamed radon in 1923. Radon has a half-life of only 3.8 days. It can be difficult for a chemist to prove he or she has discovered an element that has such a short half-life – and some elements exist for only a fraction of a second.

If finding radon was a challenge, finding francium-223 was even more so. It has the shortest half-life of any naturally occurring radioactive element – just 22 minutes. It is also the last radioactive element to have been found in nature. It was discovered in 1939 by French physicist Marguerite Perey, an assistant to Marie Curie. The last non-radioactive element found in nature was hafnium, discovered in 1923 (see page 168).

Finding francium was even harder than finding radon. It exists in uranium and thorium ores, but in very small amounts: it's estimated that only 20–30 g of francium exist in the Earth's crust at any one time. Marguerite Perey discovered it while extracting actinium from uranium ore. Like the other contaminants of uranium

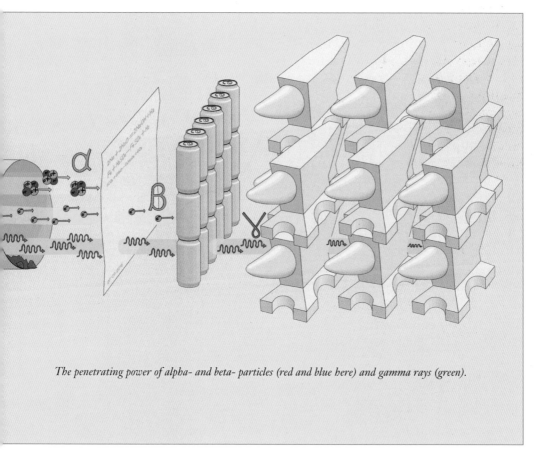

The penetrating power of alpha- and beta- particles (red and blue here) and gamma rays (green).

VIRTUALLY STABLE

Technically bismuth-209 is radioactive. But it has a half-life of 19 trillion years, decaying to thallium-205. This means that if 100 g of bismuth-209 had been present at the beginning of the universe, about 99.9999999 grams of it would still be undecayed.

as she rightly argued, the missing element 87 that Mendeléev had predicted. Other scientists disputed this, though, and her discovery of francium was not accepted until 1946.

Although many radioactive elements are formed for a longer or shorter time during a decay chain, the chain eventually stops. Sooner or later a stable element is reached. The proportion of a stable end product in a sample with a known radioactive decay chain enables scientists to calculate the age of the sample. This is the basis of carbon-dating and other dating methods that use the natural decay of radioactive isotopes as markers (see page 171).

discovered by the Curies, actinium occurs in only tiny quantities: a ton of uranium ore yields just one or two milligrams of actinium. Francium was contaminating the tiny amount of actinium that Perey extracted. After removing all the known contaminants from her sample of actinium, she found it was still too radioactive. Recognizing that there must be a different, additional source of radiation in her product, she set about isolating it. It was,

Isotopes – the same but different

Investigating radioactive decay led inevitably to the discovery of isotopes, which are variants of an element. All atoms

NO MORE STABILITY

The last of the stable elements to be discovered was the transition metal hafnium (element 72), confirmed in 1923. Every element discovered or synthesized since then has been radioactive. Using spectroscopy, Dutch physicist Dirk Coster and Hungarian radiochemist Georg von Hevesy

found hafnium in the Norwegian mineral zircon. Isolating hafnium from zirconium proved a very difficult task. Finding that most zirconium-bearing minerals also contained hafnium, Hevesy realized the atomic weight of zirconium had been wrongly reported. They produced a hafnium-free sample of zirconium for re-measurement.

Left: Hafnium, discovered in 1923, was the last stable element found.

Left: Marguerite Perey discovered francium-223, which has the shortest half-life of any naturally occurring element at just 22 minutes. It was the last radioactive element to be discovered first in nature.

of an element have the same number of electrons and protons, but they can have different numbers of neutrons in the nucleus. Isotopes were first recognized by Frederick Soddy around 1910. He realized that several substances he had found, which had different radioactive properties and different atomic weights, were in fact the same element. He gave them the name 'isotope'. Attempts to separate radioactive isotopes by chemical means always proved frustrating; inevitably so, as the chemistry of the isotopes is identical so they can't be separated in this way. Still, chemists tried for some years, unaware of the inevitability of failure. The term 'atomic weight' is used for the average of the atomic masses of isotopes

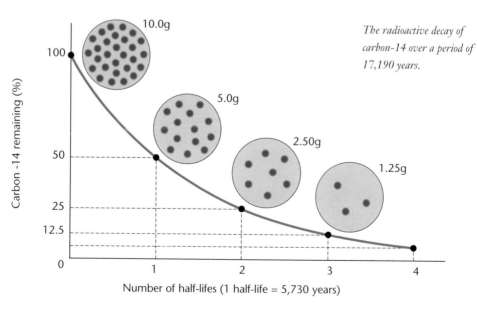

The radioactive decay of carbon-14 over a period of 17,190 years.

169

in a sample. 'Atomic mass' is used for the mass of a specific atom or isotope. So, the atomic weight of chlorine is 35.5 but that is an average of the atomic masses of the isotopes of chlorine, adjusted to reflect their relative abundance. Today, isotopes of an element are given numbers which indicate their atomic mass. For example, carbon-12 has atomic mass 12 and carbon-14 has atomic mass 14. As the neutrons carry no charge, this does not affect the electrical charge of the atom nor its chemical behaviour in terms of the ability to form bonds, but it does change its atomic mass.

Decaying in order

Soddy (and, independently, Polish chemist Kazimierz Fajans) went on to work out

---

LONG AND SHORT HALF-LIFES

The element with the longest half-life is tellurium-128, at 160 trillion times the age of the universe. If you had one gram of tellurium-128, you could expect a single atom to decay (on average) once every 674 years. So, a gram of tellurium-128 that was around at the time of the Black Death would, in all probability, have lost only one atom since then.

The element with the shortest half-life is hydrogen-7, which exists for 23 yoctoseconds, or 2.3×10^{-24} seconds. It's hard to imagine just how short a time this is. If you started counting seconds at the start of the universe, you'd only be one ten-millionth of the way to 10^{24} by now. Take that number and invert it (one over that number). That's the fraction of a second an atom of hydrogen-7 is likely to last. It was synthesized in 2003.

The half-life of bismuth-209 is 19 quintillion years (1.9×10^{19}); that is over a billion times the age of the universe.

Radioisotope	Half-life
Polonium-215	0.0018 seconds
Francium-223	22 minutes
Bismuth-212	60.5 minutes
Barium-139	86 minutes
Sodium-24	15 hours
Iodine-131	8.07 days
Cobalt-60	5.26 years
Radium-226	1,600 years
Uranium-238	4.5 billion years

> **Atomic number** = number of protons
> **Atomic mass** = number of protons + number of neutrons
> **Atomic weight** = average atomic mass of all the isotopes of an element

the law that covers radioactive decay. This makes it possible to predict which isotope will be produced by the decay of a radioactive element:

• For alpha decay, the product will have an atomic number lower by 2 and atomic mass lower by 4 than the parent isotope. For example, uranium-238 decays to thorium-234:

$$^{238}_{92}U \rightarrow {}^{234}_{90}Th + \alpha$$

• For beta decay, the product will have an atomic number higher by 1 than the parent isotope, and the atomic mass will be unchanged. For example, lead-212 decays to bismuth-212:

$$\beta + {}^{212}_{82}Pb \rightarrow {}^{212}_{83}Bi$$

How to make an element

The finding led chemists and physicists towards the discovery that they could, after all, transform one element into another. More amazing still, they could transform a known element into a totally new, unknown element, opening up a whole new chapter in the story of the Periodic Table.

Below: Carbon-14 is created naturally in the atmosphere and exists alongside the stable carbon-12. It is taken in by living things until their death. The carbon-14 decays with a half-life of 5,730 years. The proportion of carbon-14 remaining is an indication of the time that has passed since the organism died. This is the basis of carbon dating.

Going nuclear

The first nuclear reaction prompted and observed in a laboratory was the work of Rutherford in 1917. Using equipment that allowed him to bombard gases with alpha particles, he managed to 'split the atom' as it was termed at the time.

'Split' is slightly misleading; Rutherford's reaction actually added to the atom of nitrogen-14 that he started with. By smashing alpha particles into nitrogen-14 atoms, he could fuse the alpha particle to the nucleus, so increasing its count of 'nucleons' (protons and neutrons) by three, making oxygen-17, and knocking out a single hydrogen nucleus (proton):

$$^{14}N + \alpha \rightarrow {}^{17}O + p$$

FISSION AND FUSION

There are two forms of nuclear reaction: nuclear fission and nuclear fusion.

• Fission involves the depletion of the atomic nucleus by removing particles. The result of a fission reaction is atoms with a lower atomic number than the parent atom. When a large atom, such as uranium, is split, it produces two entirely different atoms. For example, uranium-235, bombarded with a neutron, produces krypton-92 and barium-141 (and some spare neutrons).

• Fusion involves adding to the atomic nucleus. The result of a fusion reaction is an atom with a higher atomic number.

Below: Particles can be accelerated to such a high speed that they collide with dramatic results.

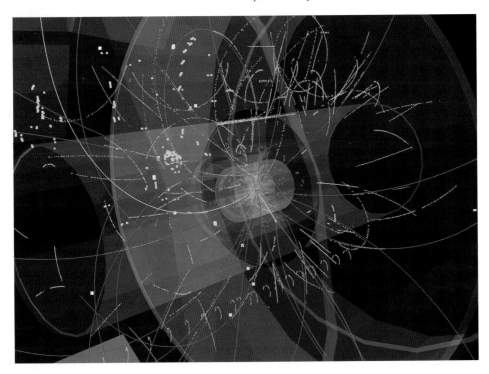

In the following years, Rutherford worked with more gases and generated more products. His findings were secure, but what he could do was limited. He had no way of giving his alpha particles sufficient energy to experiment more ambitiously, as they were simply emitted by radioactive materials in the normal course of decay. For something more dramatic, a lot more energy was needed.

Straight and curly paths

Physicists set about making particle accelerators – machines that could speed up the sluggish alpha particles so they had sufficient energy to crash into atoms and do real nuclear damage to them. The technique uses a powerful magnetic field to increase the speed of particles and spit them out as a focused, high-energy beam. This beam can then be used to bombard a substance.

The first effective accelerator was built by English physicist, John Cockcroft, and his Irish colleague, Ernest Walton. Their device achieved a nuclear reaction in 1929. Three years later, in 1932, Cockcroft and Walton bombarded lithium with protons (hydrogen ions) accelerated to 0.5 MeV, and produced alpha particles:

$$^1H + {}^7Li \rightarrow {}^4He + {}^4He$$

This was considerable progress, but Cockcroft and Walton's accelerator pushed its particles along a straight path. This meant that the speed they could achieve was limited by the length of the equipment, and it wasn't really feasible to make an extremely long accelerator.

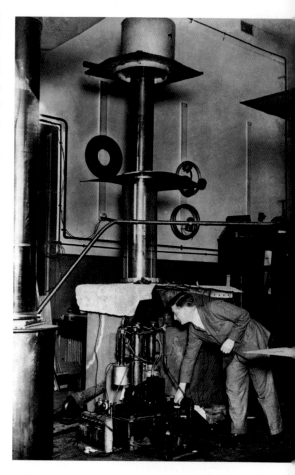

Above: John Cockcroft with his particle accelerator at the University of Cambridge, England.

The solution was found in California in 1932 (the same year as Cockcroft's disintegration of lithium) when physicist Ernest Lawrence developed the cyclotron. Instead of speeding up the particles as they travelled through a straight tube, Lawrence's equipment had them accelerating along a spiral path between two magnetizing plates. Starting at the middle of a sealed gas chamber sandwiched between two powerful magnetic coils, the particles were accelerated repeatedly by the same magnetic field; it was a far more efficient use of both space and magnetic field than a linear accelerator, and far more effective. Lawrence and his research student, Stanley Livingstone, managed to accelerate protons to one million electron volts (1 MeV), twice the speed achieved by Cockcroft and Walton with their linear accelerator.

Made by mistake

Lawrence continued his cyclotron development and experiments, and in 1937 was bombarding molybdenum (element 42) with deuterons, which are proton-neutron pairs (so, half of a helium nucleus). At this time, Italian-Jewish physicist Emilio Segrè visited Lawrence and saw a demonstration of the cyclotron. Segrè asked for some of the scrap metal that had been irradiated, which he took back to his laboratory in Sicily. Soon after, Lawrence sent him another sample, this time of molybdenum foil from

> **ELECTRON VOLTS**
>
> An electron volt is the amount of energy gained or lost by a single electron moving over a potential difference of one volt. It's approximately equivalent to 1.6×10^{-19} joules.

the deflector, which showed unexpected radioactive characteristics. It was from this foil, which had been bombarded repeatedly, that Segrè extracted technetium, named after the Greek *teknetos*, 'man-made'.

While Segrè was on another visit to Berkeley in 1938, to work on short-lived isotopes of technetium, the fascist dictator Benito Mussolini passed laws barring Jews from holding academic office in Italy. Segrè decided to remain in the USA, and worked with Lawrence as a research assistant. Segrè discovered a second element, astatine, in 1940. The name is from the Greek for 'unstable'. Even the most stable isotope of astatine has a half-life of only 8.1 hours.

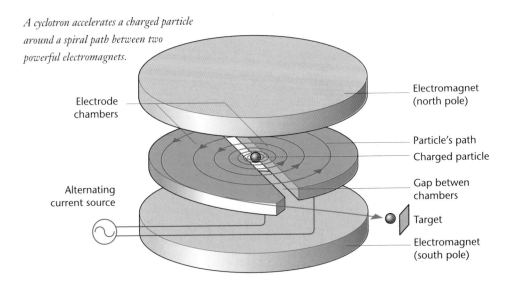

A cyclotron accelerates a charged particle around a spiral path between two powerful electromagnets.

Electrode chambers

Alternating current source

Electromagnet (north pole)

Particle's path

Charged particle

Gap betwen chambers

Target

Electromagnet (south pole)

Segrè turned his attention to the last gap for a non-transuranic element (atomic number less than 92) in the Periodic Table, element 61, but was unable to separate it.

Discovered twice

It looks likely that technetium was discovered in nature before it was made in Lawrence's cyclotron. In 1925, geochemists Ida Noddack-Tacke, Walter Noddack and Otto Berg published a paper announcing the discovery of an element they named 'masurium' after the area, now in Poland, where Walter's family originated. They claimed rhenium in the same article. They could not isolate sufficient 'masurium' to

The enormous hadron collider built beneath the France–Switzerland border near Geneva, is the descendant of Lawrence's cyclotron. Accelerated around a circular path, particles can collide at speeds and levels of energy unimagined by the early nuclear physicists.

support their claim and it was rejected by the scientific community.

However, when research chemist David Curtis repeated the experiment in 1999, he found it matched the proportions of naturally occurring technetium that could be expected in the starting material, columbite. Columbite can contain up to 10 per cent uranium, and uranium can contain

---

UN-TECHNETIUM

Element 43 was the last of Mendeléev's predicted elements to be discovered. Eka-manganese, as he had named it, was wrongly thought to have been found at least once by the time Segrè identified it. The first announcement of its discovery was in 1908, by Japanese chemist Masataka Ogawa. He named it 'nipponium' after the Japanese name for Japan, but his new element turned out to be rhenium. Unfortunately, he didn't get credit for rhenium either, which was officially discovered in 1925 (see above).

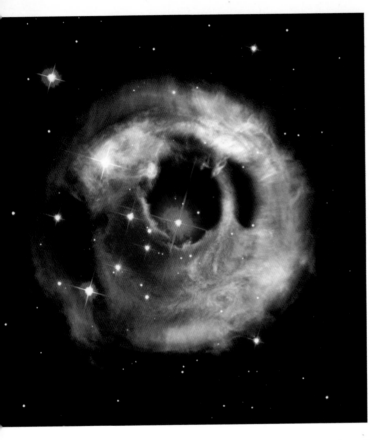

Red giant stars produce heavy elements at their core as they near the end of their life.

4.2 million years. So surely there should not be any technetium left in these ancient and decaying stars? The only explanation could be that the stars are themselves producing technetium. And this was the clue to where all the elements come from.

Further steps in messing with atoms

Instead of trying to make new elements, the Joliot–Curies put particle bombardment to a slightly different use, discovering induced radioactivity in 1933. They found that if they bombarded light elements such as boron and aluminium (which are not normally radioactive) with alpha particles, the atoms would emit radioactive radiation – and this would continue even after the bombardment stopped. The radiation emitted turned out to be a form of beta particle: a positively-charged particle with the mass of an electron (a positron). The fact that the production of radium continued after bombardment ceased showed they had found a way of inducing normally stable isotopes to become radioactive. They had successfully made radioactive phosphorus-30 from inert

about 1 mg of technetium per kilogram. It appears that the Noddack–Tackes experiment had discovered 'masurium', twelve years before Segrè found technetium in waste molybdenum.

Lost in space

Technetium had another surprise to deliver. In 1952, Paul Merrill, an astronomer at the Mount Wilson and Palomar observatories in California, noticed that the spectra of red giants (stars near the end of their lifespan) showed these stars to be rich in technetium. This was something of a paradox, as the longest half-life of any technetium isotope is aluminium-27. The immediate application of their discovery was the use of radioactive tracers to reveal biological processes and the sequence of other complex reaction chains. By making an innocent element radioactive, its activity could be tracked. For radiochemists, it had other uses.

Struggling to get beyond uranium

All the elements discovered since 1940 have been radioactive, created in nuclear reactions using particle accelerators. If any of these have ever been created in nature in the heart of a supernova, they have not made it into the composition of the planets and stars that scientists have examined. Most have a very short half-life – and very short in this case can mean just seconds, milliseconds or even less. Therefore, although they might form in supernovae, they won't survive long.

Until 1932 and the discovery of the neutron, no one really gave much thought to the possibility of elements beyond uranium – labelled transuranic elements. But the neutron provided another bit of ammunition for firing at atomic nuclei. Transuranic elements became a real possibility and, soon, a preoccupation of several radiochemists.

Neutrons last just a few minutes outside an atomic nucleus, so neutron-bombardment can only be achieved by disintegrating atoms in a nuclear reaction or in particle accelerators and then whizzing the escaping neutrons up to speed before they decay. Both of these methods would become the focus of research and provide the path to new elements.

Below: **Blue:** *Elements with at least one stable isotope;* **Green:** *Most stable isotope has a half-life of over two million years;* **Yellow:** *Most stable isotope has a half-life of 800–34,000 years;* **Orange:** *Most stable isotope has a half-life of 1 day–103 years;* **Red:** *Most stable isotope has a half-life of less than one day;* **Purple:** *Most stable isotope has a half-life of less than several minutes.*

1 H																	2 He
3 Li	4 Be											5 B	6 C	7 N	8 O	9 F	10 Ne
11 Na	12 Mg											13 Al	14 Si	15 P	16 S	17 Cl	18 Ar
19 K	20 Ca	21 Sc	22 Ti	23 V	24 Cr	25 Mn	26 Fe	27 Co	28 Ni	29 Cu	30 Zn	31 Ga	32 Ge	33 As	34 Se	35 Br	36 Kr
37 Rb	38 Sr	39 Y	40 Zr	41 Nb	42 Mo	43 Tc	44 Ru	45 Rh	46 Pd	47 Ag	48 Cd	49 In	50 Sn	51 Sb	52 Te	53 I	54 Xe
55 Cs	56 Ba	*	72 Hf	73 Ta	74 W	75 Re	76 Os	77 Ir	78 Pt	79 Au	80 Hg	81 Tl	82 Pb	83 Bi	84 Po	85 At	86 Rn
87 Fr	88 Ra	**	104 Rf	105 Db	106 Sg	107 Bh	108 Hs	109 Mt	110 Ds	111 Rg	112 Cn	113 Nh	114 Fl	115 Mc	116 Lv	117 Ts	118 Og

*Lanthanides		57 La	58 Ce	59 Pr	60 Nd	61 Pm	62 Sm	63 Eu	64 Gd	65 Tb	66 Dy	67 Ho	68 Er	69 Tm	70 Yb	71 Lu
**Actinides		89 c	90 Th	91 Pa	92 U	93 Np	94 Pu	95 Am	96 Cm	97 Bk	98 Cf	99 Es	100 Fm	101 Md	102 No	103 Lr

Neutrons to the rescue

Italian chemist Enrico Fermi realized that although the Joliot–Curies had used alpha particles, bombarding atoms with the newly discovered neutrons might produce better results. As protons have a positive charge, the nucleus of the target atom repels them, but it does not repel the chargeless (neutral) neutron.

In 1934, Fermi started to bombard inert elements with neutrons to see what would happen. He found that lighter elements would emit a proton (so their atomic number decreased) but heavier elements would absorb a slow-moving neutron, converting

it to a proton. They would therefore put on weight and move further along the table. Fast-moving neutrons tended to knock particles out of the nucleus.

Transuranics under scrutiny

One product of Fermi's investigations was an isotope with a half-life that he couldn't match to any known element. Had he made element 93 by forcing an extra neutron into uranium, where it would become a proton and an electron? When he cautiously published the finding in 1934 there were plenty of objections from his contemporaries. He certainly had products with half-lifes that didn't match known isotopes, but could not isolate the possible element 93, now known as neptunium. The confusion lasted for years.

Then, a month after Fermi had won a Nobel Prize in 1938 for demonstrating 'the existence of new radioactive elements produced by neutron irradiation', nuclear fission was achieved and his unidentified element was shown to be a fission product. This possibility had been suggested by Ida Noddack, but ignored because fission was not considered possible.

Left: Enrico Fermi in his laboratory.

WAS 93 THERE ALL ALONG?

Among those attempting and claiming to find element 93 (neptunium) were Romanian physicist Horia Hulubei and French chemist Yvette Cauchois. In 1938, while using spectroscopy to examine minerals, they found a new element which they named 'sequanium'. Their claim was disregarded on the grounds that element 93 was not thought to be naturally occurring. Neptunium is, in fact, found in small quantities in uranium ore, so it remains possible that they really did find it.

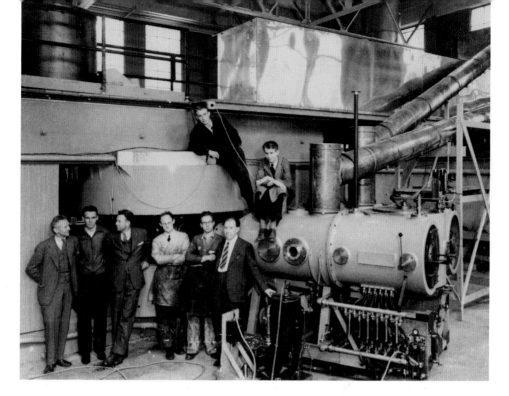

Getting it wrong and getting it right

While Fermi thought he had found the first transuranic element when he hadn't, Edwin McMillan concluded that he hadn't found a transuranic element when in fact he had. Using a new 60-inch cyclotron at Berkeley, California, McMillan bombarded uranium with the aim of examining the separated fission products. He found two new half-lifes in the uranium trioxide he was using. One was recognizable as an isotope of uranium (uranium-239) but the other, with a half-life of 2.3 days, was unknown. He set about investigating whether it could be element 93. At the time, people expected element 93 to behave like rhenium, but his product did not.

The following year, working with Philip Abelson, McMillan recognized the product as more like a rare-earth element and successfully isolated it. The two men produced a larger sample and showed

Above: The 60-inch cyclotron at the Lawrence Radiation Laboratory in Berkeley, California, soon after its completion in 1939. Lawrence is third from the left; he won the Nobel Prize in Physics for his invention.

conclusively that the product with the 2.3-day half-life increased in abundance as the uranium-239 decreased. Clearly, the uranium was changing into the new element, which was then decaying into something else. They had found not only element 93, but also element 94, which it decayed into. They could not isolate enough of element 94 to prove its chemistry, however.

The discovery of element 94 is credited to American chemist Glenn Seaborg (1912–99), who identified plutonium-238 later the same year. Neptunium, as it was named, could only be properly explored after 1942, when Seaborg and Arthur Wahl isolated a different isotope, neptunium-237, which has a half-life of 3 million years.

NEW PLANET, NEW ELEMENTS

Element 93 was named neptunium after the planet Neptune (the planet in the solar system after Uranus) because it comes after uranium in the Periodic Table. Plutonium was named after Pluto, which had been discovered in 1930 beyond the orbit of Neptune and was still considered a planet in 1940 (it's now known as a 'dwarf planet'). If planetary naming had gone differently, we might have differently named elements. William Herschel wanted to name Uranus after King George III of England. Pluto was named by an 11-year-old girl, Venetia Burney, after the Greek god of the underworld. Neither she nor Herschel could have imagined they would get an element thrown in for free!

Methods of making

The methods of natural synthesis run out around plutonium (element 94). It's thought there's only about 0.05 g of naturally occurring plutonium in the world at any one time, so the chances of stumbling across it are pretty small. Elements beyond plutonium only exist – and possibly have only ever existed – in the laboratories of 20th- and 21st-century physicists.

The hunt is on

The search for transuranic elements began in earnest after World War II. During the war, research effort was focused on plutonium, which could be used in a nuclear weapon. The process of

Glenn Seaborg indicating seaborgium, the element named after him, in the Periodic Table.

firing high-energy neutrons at uranium and waiting for the inevitable radioactive decay to do its magic had produced neptunium and plutonium, and soon produced more. Seaborg and others produced americium in 1944 as part of the Manhattan Project, but it was kept secret until after the war. Americium turned up mixed with curium (another element that was kept secret) and these two proved so difficult to separate that the team privately referred to them as 'pandemonium' and 'delirium'.

In the end, curium (element 96) was isolated first and named after Marie Curie. Americium (element 95) soon followed and was named after America because its position in Seaborg's newly adapted Periodic Table fell immediately below europium. Now, most americium and curium is produced by bombarding uranium or plutonium with neutrons in nuclear reactors. A tonne of spent reactor fuel yields just 100 g of americium and 20 g of curium. Americium is widely used in smoke detectors, while curium is used to power satellites. Understandably, there's a far larger market for americium than for curium. The heavier elements are probably non-existent in nature (though there might be a little americium occasionally).

Seaborg and the Periodic Table

Seaborg and his colleagues went on to discover (or, rather, make) a total of 11 new elements. But it turned out that they could not all be tacked on to the end of the Periodic Table. Just as the transition metals had prised apart Mendeléev's equal columns and the lanthanides had disrupted the transition metals (see page 182), the actinides would need a special place. In 1944, Seaborg had formulated the 'actinide concept', predicting that the actinides would form a transition series like the lanthanides. The series runs from actinium (element 89), which comes straight after radium and gives the series its name, to lawrencium (element 103).

Below: Americium in smoke detectors emits alpha particles which ionize air, causing a flow of particles between positive and negative plates. The alarm triggers when smoke disrupts the flow of particles.

THE UNWANTED PRODUCT

After all the effort spent looking for element 93, it's rather disappointing that most research since has been into methods of getting rid of it. Neptunium-237 is considered an unwanted by-product that presents particular problems of containment. With a half-life now established at 2.14 million years, it has prompted designs for improved waste containment facilities that will last for thousands of years.

Chemists had assumed the lanthanides (elements 57–71) were an aberration and that the periodic pattern would continue uninterrupted after them. Only Seaborg anticipated that they introduce a new aspect of periodicity. If the actinides didn't have a series of their own, but continued sequentially through the table from actinium, then neptunium (element 93) would come directly below rhenium. The assumption that it would, and the expectation that it would share properties with rhenium, led to confusion when Fermi thought he had found element 93 but was perturbed that it was nothing like rhenium.

Rather than trying to shoehorn the new elements into the existing table, Seaborg proposed a redesign which has now become standard. The lanthanides and the actinides appear in their own rows beneath the table, with a keyline showing where they should be inserted. It's not the only way of representing their position, but it is now the most widely used. Seaborg went on to predict the existence of a transactinide series and even a superactinide series. The first would hold elements 104–121, and the second would continue the table from 122–153. There is so far no evidence that any superactinide elements exist.

Orbitals and periodicity

The number of elements in each part of the Periodic Table reflects the arrangement of electrons within the atoms. As we have seen (page 156), the innermost shell has room for only two electrons. Hydrogen has one electron here and helium has two. Then it's full, so the next atom, lithium, has to start a new shell. This has space for eight

Below: The Periodic Table, drawn to include the lanthanides and actinides incorporated in the right places, is so wide that it's cumbersome.

s	d					g																						

IA
1	H 1	IIA																						
2	Li 3	Be 4																						
3	Na 11	Mg 12	IIIB																					
4	K 19	Ca 20	Sc 21																					
5	Rb 37	Sr 38	Y 39															Ce 58	Pr 59	Nd 60	Pm 61
6	Cs 55	Ba 56	La 57															Th 90	Pa 91	U 92	Np 93
7	Fr 87	Ra 88	Ac 89																		

Row 6 (right block): Ce 58, Pr 59, Nd 60, Pm 61
Row 7 (right block): Th 90, Pa 91, U 92, Np 93

Row 8:
Uue 119	Ubn 120	Ubu 121	Ubb 122	Ubt 123	Ubq 124	Ubp 125	Ubh 126	Ubs 127	Ubo 128	Ube 129	Utn 130	Utu 131	Utb 132	Utt 133	Utq 134	Utp 135	Uth 136	Uts 137	Uto 138	Ute 139	Uqn 140	Uqu 141	Uqb 142	Uqt 143

Row 9:
Uhs 169	Usn 170	Usu 171	Usb 172	Ust 173

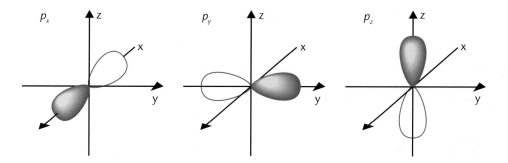

electrons, arranged in four pairs. Each pair is designated by a letter that indicates the shape of the orbital. In level n=1, the single pair of electrons occupies spherical orbits, indicated by 's'. The electron configuration of hydrogen is given as $1s^1$, and helium is $1s^2$. Level n=2 has an 's' orbital that can hold two electrons, and three 'p' orbitals that are roughly hemispherical or lobe-shaped, and again each holds up to two electrons. They

The three p-orbitals are oriented along perpendicular axes so that the electrons are as widely separated as possible, with two lobe-shaped orbitals in each one.

are differently oriented so that the electrons can keep as far away from one another as possible. To distinguish between the p-orbitals they are given subscript letters: p_x, p_y and p_z.

f									d								p						

																							0
																							He
																IIIA	IVA	VA	VIA	VIIA			2
																B	C	N	O	F			Ne
																5	6	7	8	9			10
																Al	Si	P	S	Cl			Ar
																13	14	15	16	17			18
							IVB	VB	VIB	VIIB		VIII		IB	IIB								
							Ti	V	Cr	Mn	Fe	Co	Ni	Cu	Zn	Ga	Ge	As	Se	Br			Kr
							22	23	24	25	26	27	28	29	30	31	32	33	34	35			36
							Zr	Nb	Mo	Tc	Ru	Rh	Pd	Ag	Cd	In	Sn	Sb	Te	I			Xe
							40	41	42	43	44	45	46	47	48	49	50	51	52	53			54
Sm	Eu	Gd	Tb	Dy	Ho	Er	Tm	Yb	Lu	Hf	Ta	W	Re	Os	Ir	Pt	Au	Hg	Tl	Pb	Bi	Po	At	Rn
62	63	64	65	66	67	68	69	70	71	72	73	74	75	76	77	78	79	80	81	82	83	84	85	86
Pu	Am	Cm	Bk	Cf	Es	Fm	Md	No	Lr	Ku	Ha	Sg	Ns	Hs	Mt	Ds	Rg	Cn	Nh	Fl	Mc	Lv	Ts	Og
94	95	96	97	98	99	100	101	102	103	104	105	106	107	108	109	110	111	112	113	114	115	116	117	118
Uqq	Uqp	Uqh	Uqs	Uqo	Uqe	Upn	Upu	Upb	Upt	Upq	Upp	Uph	Ups	Upo	Upe	Uhn	Uhu	Uhb	Uht	Uhq	Uhp	Uhh	Uhs	Uho
144	145	146	147	148	149	150	151	152	153	154	155	156	157	158	159	160	161	162	163	164	165	166	167	168

Each orbital first gets one electron; when all orbitals have one they start to gain another, working from the orbital closest to the nucleus outwards. Looking at the second period, the electron configurations of the elements are:

Li	Be	B	C	N	O	F	Ne
3	4	5	6	7	8	9	10
$1s^2$	$1s^2$	$1s^2$	$1s^2$	$1s^2$	$1s^2$	$1s^2$	$1s^2$
$2s^1$	$2s^1$	$2s^1$	$2s^1$	$2s^2$	$2s^2$	$2s^2$	$2s^2$
		$2p_x^1$	$2p_x^1\ 2p_y^1$	$2p_x^1$	$2p_x^2\ 2p_y^1$	$2p_x^2\ 2p_y^2$	$2p_x^2\ 2p_y^2$
			$2p_z^1$	$2p_y^1$	$2p_z^1$	$2p_z^1$	$2p_z^2$

Beyond these, the next shell has 18 orbitals, the next has 32, and the next has 50. The letters for subshells are repeated in each shell:

Shell	Maximum electrons	Subshells
1	2	s2
2	8	s2, p6
3	18	s2, p6, d10
4	32	s2, p6, d10, f14
5	50	s2, p6, d10, f14, g18

The Periodic Table is often shown with the elements colour-coded and labelled in different blocks which are designated by letters – the same letters as are used for the outer occupied orbitals of the atoms. The transition metals appear as soon as d-orbitals begin to appear in the atoms. The appearance of the lanthanide and actinide sequences corresponds to that of the f-orbitals. Seaborg's recognition that the actinides are a continuation rather than an anomaly of periodicity becomes clear when the electron configurations are considered.

Exploding into life

The method used by Seaborg and others to synthesize new elements worked well enough up to fermium (element 100), but once the nucleus contains a very large number of protons they repel each other strongly enough to make it difficult to add any more and get them to stick. Indeed, even fermium and einsteinium (element 99) were created inadvertently as by-products of the testing of the first hydrogen bomb in 1952.

The big guns

Scientists switched from trying to insert individual neutrons into an atomic nucleus to trying instead to fuse two completely different nuclei together. The first success came in 1955, fusing the newly found einsteinium with helium to make mendelevium (element 101). Others followed from the 1960s onwards.

Only moscovium (element 115) was produced by a different method. Synthesized in 2010 in Russia, it was made by the alpha-particle decay of tennessine (element 117). Clearly it could only be discovered after tennessine had been synthesized and had had time to decay.

Left: The Periodic Table, showing the configuration of electrons in the outermost occupied shell of the atoms of each element.

185

The end is in sight

For now, chemistry stops at oganesson (element 118). Curiously, Niels Bohr predicted the possibility of oganesson in 1922. It makes a natural neat end, as it is at the bottom of Group 18, the noble gases, so there are no unfilled slots if this is indeed the final element.

Not everyone is confident there will be no more elements. The American physicist Richard Feynman predicted that elements might be synthesized up to 137. Seaborg allowed for superactinides up to element 153, and Finnish chemist Pekka Pyykkö allowed elements up to 172 in his extended Periodic Table.

103	Lawrencium	1961	Bombardment of californium with boron
102	Nobelium	1966	Bombardment of uranium with neon
104	Rutherfordium	1969	Bombardment of californium with carbon
105	Dubnium	1970	Bombardment of californium with nitrogen
106	Seaborgium	1974	Bombardment of californium with oxygen
107	Bohrium	1981	Bombardment of bismuth with chromium
109	Meitnerium	1982	Bombardment of bismuth with iron
108	Hassium	1984	Bombardment of lead with iron
110	Darmstadtium	1995	Bombardment of lead with nickel
111	Roentgenium	1995	Bombardment of bismuth with nickel
112	Copernicium	1996	Bombardment of lead with zinc
113	Nihonium	2004	Bombardment of bismuth with zinc
114	Flerovium	2004	Bombardment of plutonium with calcium
116	Livermorium	2004	Bombardment of curium with calcium
118	Oganesson	2006	Bombardment of californium with calcium
117	Tennessine	2010	Bombardment of berkelium with calcium

THE LEGENDARY ISLAND OF STABILITY

The heaviest metal to have been investigated is flerovium, element 114. It's not clear whether it is closer in chemistry to a heavy metal or to a gas. It's very volatile and unreactive. Only 90 atoms have been observed, and with a half-life of just 2.6 seconds there is little time to find out much about those atoms before they're gone. An isotope with a longer half-life of 19 seconds, flerovium-290, might exist but is unconfirmed. It has an unusually long half-life for such a super-heavy element, and this is sometimes considered evidence of a fabled 'island of stability'. This is the theoretical existence of stable isotopes of some elements with very large atomic numbers, beyond the elements so far discovered.

Most of the very heavy elements have extremely short half-lifes because their elements have so many protons that they are forcing themselves apart. But some scientists think there might be 'magic numbers' of protons and neutrons that put themselves into more stable arrangements. (Just as the electrons are arranged into shells and orbitals, the protons and neutrons are similarly arranged inside the nucleus.) The idea, suggested by Glenn Seaborg, is supported by elements 110–114 being more stable than expected, and even the one-millisecond half-life of element 118 is longer than predicted. A 'magic number' of 184 neutrons is considered likely to produce a more stable element. If larger nuclei are indeed stable, it is even possible that some further elements occur in nature (though perhaps not on Earth).

The ability of radioactive sources to produce energy at a predictable rate, regardless of environmental conditions and without recourse to other chemicals, makes them an excellent power source for spacecraft. NASA's New Horizons probe sent to investigate Pluto is powered by the radioactive decay of 11 kg of plutonium dioxide.

Having discovered they can make atoms from others, why would chemists stop in their quest for ever-heavier elements? One reason is that the half-life of the heavy elements becomes very short and they are incredibly difficult (and expensive) to create. Enough weapons-grade plutonium to make a viable bomb (about 6 kg) would cost in the region of $30 million. Californium (element 98) is the heaviest element produced in any kind of commercial quantity and costs $66 million a gram. The heaviest element, oganesson, has a half-life of less than a millisecond and only five (or possibly six) atoms have ever been created.

It takes years for an expert research team to make a few atoms that are gone in a thousandth of a second. But the techniques used to produce those few atoms were undreamt of when Bohr first suggested that element 118 might be found. Who is to say what techniques might yet lie ahead of us? Polish nuclear physicist Witold Nazarewicz suggests there might be around 7,000 possible 'nuclides' (configurations of nuclei, or combinations of protons and neutrons). Beyond that, others would have half-lifes so short they would never form. Currently, the 118 elements are known to exist in a total of about 3,000 nuclides.

> 'We must be clear that when it comes to atoms, language can be used only as in poetry. The poet, too, is not nearly so concerned with describing facts as with creating images and establishing mental connections.'
>
> Niels Bohr, 1920

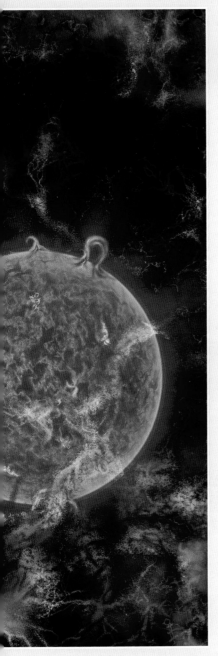

CHAPTER 9

CELESTIAL ELEMENT FACTORIES

'The nitrogen in our DNA, the calcium in our teeth, the iron in our blood, the carbon in our apple pies were made in the interiors of collapsing stars. We are made of star stuff.'

Carl Sagan

One of the most astonishing, inspiring and poetic discoveries of the 20th century is that all the naturally occurring elements in our world were forged in the hearts of long-dead stars.

In the early universe, blue supergiant stars were the first factories producing the heavier elements. Signals left over from these earliest stars were detected in 2018.

The gods and the stars

For the Ancient Greeks and many other old civilizations, the origin of all matter was primeval chaos, which could be a void or a state of formless matter. In the West, the notion that the universe was created by a divine being remained pretty much unchallenged, at least overtly, until at least the 18th century. Eventually, however, scientific enquiry led in a direction that

The age of the Earth demands a Sun that will produce energy for at least as long as the Earth has already lasted – a puzzle for 19th-century scientists.

would elucidate the origins of the elements. It arose out of questions concerning the age of the Earth, an early point of conflict between Creationism and science.

What makes the Sun shine?

At the beginning of the 20th century, the energy that powers the Sun was generally thought to be produced by gravitational contraction or the shrinking Sun model (see box). This had been suggested by German physicist Hermann von Helmholtz in 1850. The theory served while estimates for the age of the Earth were wildly short of reality, but in the late 19th century it emerged that the Earth was almost certainly much older than previously thought. Lord Kelvin calculated that a large Sun contracting could only provide energy of the quantities needed for around 35 million years. The Earth, geologists were now claiming, was much older than that. By the end of the century, most of them agreed on a figure of around 100 million years, but some even suggested 2,000 million years. This falls far short of the current figure of around 4,550 million years, but it still conflicted with Kelvin's theory of a star that would last a mere 35 million years.

The geologists remained at loggerheads with Helmholtz and Kelvin on this subject. This impasse was not breached until

> **HOW A SHRINKING SUN MAKES HEAT**
> The shrinking Sun model explains the Sun's production of heat and light in terms of this cycle:
> • beginning in hydrostatic equilibrium, the internal pressure of the Sun and its gravity are balanced
> • some of the Sun's energy leaks away in luminosity (it is lost as sunlight)
> • loss of energy reduces the heat and pressure inside the Sun and gravity's force becomes greater than the force of internal pressure
> • gravity causes the Sun to contract, raising the internal pressure until it is in equilibrium again
> • the cycle starts again.

new methods of dating the Earth (using radioactivity) became available and showed that the calculations of Helmholtz and Kelvin were wrong. Evidently something else was enabling the Sun to produce enough energy to shine for billions of years.

Helium and hydrogen

Arthur Eddington, an English astronomer who had provided proof of Einstein's theory of relativity in 1919, made the insightful

> 'Sir Ernest Rutherford has recently been breaking down the atoms of oxygen and nitrogen, driving out an isotope of helium from them; and what is possible in the Cavendish laboratory may not be too difficult in the Sun.'
> Arthur Eddington, *The Internal Constitution of the Stars*, 1926

suggestion in 1920 that Einstein's equation could explain where stars get their power. The equation $E = mc^2$ states that energy and mass are interchangeable:

$$\text{energy} = \text{mass} \times (\text{speed of light})^2$$

In *The Internal Constitution of the Stars*, Eddington proposed that the Sun (and other stars) could be powered by the conversion of matter into energy. English chemist and physicist Francis Aston had already demonstrated that the mass of helium is 0.8

per cent less than the mass of four hydrogen atoms. This suggests that if four hydrogen atoms can be fused to create a helium atom, there will be a little bit of matter (0.8 per cent of the mass of the starting material) that is converted into energy.

Support for Eddington's view came from the PhD thesis of a young English astronomer called Celia Payne. In 1925 she published the conclusion of her work in stellar spectroscopy: that stars are made mostly of hydrogen and helium. This discovery, combined with Einstein's special-relativity equation (1905) and the recognition around 1920 that the proton and the hydrogen nucleus are one and the same, meant that all the pieces were in place for the power behind sunshine to be revealed.

Inside the nuclear furnace

German nuclear physicist Hans Bethe finally put the pieces together. Like many scientists, Bethe left Germany in 1933 with the rise of the Nazis and made his way to the USA. There he worked on nuclear fission and fusion, becoming a reluctant participant in work on the hydrogen bomb. (He hoped that his part would be to prove it was impossible to make one.) He worked on the processes taking place inside stars, investigating their production of energy, and, it turned out, of the elements. He showed conclusively in 1938 that nuclear fusion powers the Sun and other stars, by producing energy from the fusion of four atoms of hydrogen into one atom of helium. He identified the chain of reactions which produces this result, using

Hans Bethe finally solved the problem of exactly how the Sun generates energy from matter.

carbon as a catalyst. This is known as the CNO (carbon–nitrogen–oxygen) cycle. The reaction sequence is as follows:

Step 1. $^{12}C + {}^1H \rightarrow {}^{13}N + \gamma$
Step 2. $^{13}N \rightarrow {}^{13}C + e^+ + \nu_e$
Step 3. $^{13}C + {}^1H \rightarrow {}^{14}Nc + \gamma$
Step 4. $^{14}N + {}^1H \rightarrow {}^{15}O + \gamma$
Step 5. $^{15}O + {}^{15}N + e^+ + \nu_e$
Step 6. $^{15}N + {}^1H \rightarrow {}^{12}C + {}^4He$

In the course of the reaction, carbon-12 is converted first to nitrogen-13, then to carbon-13, then nitrogen-14, then oxygen-15, then nitrogen-15, and finally reverts to carbon-12. In the process it accumulates and finally relinquishes the subatomic particles needed to build helium from hydrogen.

Of course, this begs the question of where hydrogen and helium come from – not to mention the carbon-12. But that wasn't Bethe's first concern; he was trying to find the source of the stars' energy, and he had succeeded. Calculating from the quantity of hydrogen in the Sun (35 per cent by mass), Bethe concluded that it would have sufficient fuel to keep producing energy for around 35 billion years (in reality it will only last another 5 billion years or so).

First things first

The question of where the hydrogen and helium came from was addressed soon after by Ralph Alpher and his PhD supervisor George Gamow. They published a paper in 1948 with the title 'The Origin of Chemical Elements' in which they argued that all the known elements could have come into being very soon after the Big Bang. (Big

Belgian priest Georges Lemaître was the first scientist to propose a theory of a universe expanding from an infinitesimal point.

Bang theory emerged in 1927, proposed by Belgian priest and amateur astronomer Georges Lemaître.)

Alpher and Gamow started with the concept of an early universe awash with a highly compressed 'soup' of neutrons. As the universe expanded, some of the neutrons, they proposed, decayed to produce a proton and an electron. The first stage of synthesis had a neutron and a proton colliding and combining, making a deuterium nucleus. (Deuterium is an isotope of hydrogen, hydrogen-2, that has a neutron in addition to the usual proton.) Alpher and Gamow said that to create heavier nuclei all that had to occur was the capture of another neutron or proton (nucleon), and another, and so on. The incremental addition of more and more nucleons could produce all the other

ELEMENT FACTORIES

In the first three or four minutes after the Big Bang, hydrogen nuclei formed and large numbers underwent fusion, creating helium as well as much smaller quantities of other substances including deuterium (hydrogen-2), tritium (hydrogen-3), lithium and beryllium. All this happened in the first 20 minutes of the life of the universe, stopping when its expansion meant that conditions were no longer suitable. The universe had already become too cool (at a billion degrees Celsius) and too diffuse.

As the early phase of the universe was characterized by a high-energy soup of fast moving neutrons, this result is not surprising. When neutrons decay, they convert to a proton, releasing an electron of energy. When they collide and stick together, one neutron converts to a proton. The proton-neutron pair is hydrogen-2, or deuterium. If it collides with another proton, it becomes hydrogen-3, or tritium. One more, and it has two protons and two neutrons, becoming a helium nucleus. Further collisions with helium produced lithium (element 3) and beryllium (element 4), which existed in much smaller quantities, and boron (element 5), which existed in tiny quantities. These particles were not yet atoms, in the early minutes of the universe, they were only nuclei. It would take another 380,000 years for the universe to become sufficiently cool and calm for atoms to form without instantly ripping apart again.

Below: The evolution of the universe following the Big Bang.

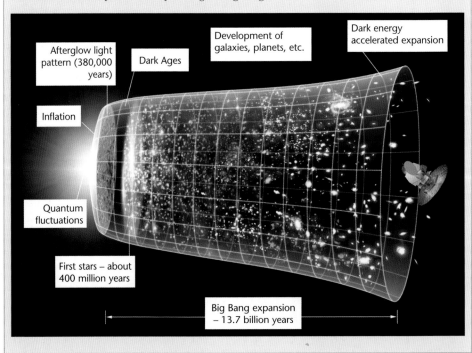

A computer model of the process of star formation from the collapse of gas clouds in the early universe when only a very limited number of elements existed.

elements and isotopes. *The Washington Post* ran a story on the theory that announced 'World Began in 5 Minutes, New Theory.'

While this mechanism works as far as the creation of helium, it can go no further. There is no element with five nucleons, so no stepping-stone from helium to elements with higher atomic numbers. Soon afterwards, it was shown that nor is there a stable isotope with atomic mass 8. Even so,

ALPHA, BETA, GAMMA

Alpher and Gamow co-opted Bethe as a sleeping-partner author of their seminal paper in 1948 so that they could list the contributors as Alpher, Bethe, Gamow – a play on alpha, beta, gamma, the first three letters of the Greek alphabet.

Alpher and Gamow's theory explained the 99 per cent of the matter in the universe that comprises hydrogen and helium.

Once Bethe had demonstrated that helium could be built from hydrogen, it became readily apparent that anything else could also be made of the same subatomic building blocks. But it wasn't possible to build the other elements by slowly pushing up the number of nucleons, as Alpher and Gamow had proposed.

The CNO cycle identified by Bethe in 1938 can clearly only work in an environment where there is carbon-12 available to start the process. There was no carbon-12 in the early universe, when the vast majority of helium was formed following the process set out by Alpher and Gamow. So what happened in between?

Forged in the stars

About 180 million years after the Big Bang, stars began to form from collapsing clouds of primordial matter. The universe contained a great deal of hydrogen and helium, and small quantities of lithium, beryllium and boron – there was nothing heavier. Again, Eddington made a remarkably prescient suggestion in 1920:

'The position may be summarized in these terms: the atoms of all elements are built of hydrogen atoms bound together, and presumably have at one time been formed from hydrogen; the interior of a star seems as likely a place as any for the evolution to have occurred; whenever it did occur a great amount of energy is being set free which is hitherto unaccounted for. You may draw a conclusion if you like.'

Eddington's suggestion was soon expanded into a thorough-going theory. In 1946, English astronomer Fred Hoyle proposed that stars might prove to be the breeding ground for the elements. Hoyle was one of three astronomers responsible for working out the conditions inside a red giant, so the idea now had a solid basis. In 1951, Austrian-born American

Astronomer and scientist Sir Fred Hoyle, pictured in his home town of Bournemouth, England, in 1994. Hoyle suggested that life, or the building blocks of life, could be carried to planets by comets or drifting interstellar dust particles.

HOYLE AND THE BIG BANG

Oddly, Fred Hoyle rejected Big Bang theory. (It was he who came up with the name 'Big Bang', using the term derisively.) Instead, he argued that hydrogen is continuously created from the vacuum of space and energy, with no need for any origin – it was spontaneous creation *ex nihilo*. Creation from the void and chaos – again!

astrophysicist Edwin Salpeter showed that three helium nuclei (atomic mass 4) could probably be fused into a single carbon nucleus (atomic mass 12) in the heart of a red giant star – but not in the Big Bang.

In 1952, the first irrefutable evidence of nucleosynthesis at work in stars arrived in the form of spectroscopic analysis. This showed the presence of technetium, that first synthetic element, in the atmosphere

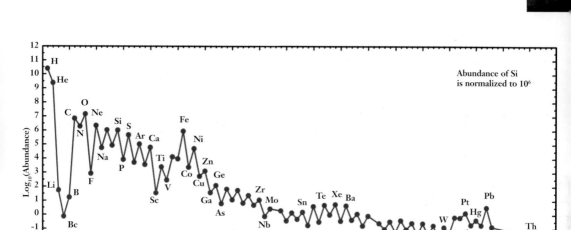

The Suess/Urey graph of the relative abundance of the elements in the solar system.

of red giants. As technetium has a maximum half-life of 4.2 million years, and the stars are billions of years old, it must have been produced in the star, probably during day-to-day activity. It could not have been there all along.

What's there?

In 1957, Austrian physicist and chemist Hans Suess and American chemist Harold Urey drew up a table of the abundance of the elements plotted against atomic number (see above). As the abundance differs dramatically, they had to use a logarithmic scale to get all the figures on the same graph. Although the relative abundances are not instantly obvious at a glance, something else is. Abundance follows a general trend of reducing as atomic number increases, which is to be expected as the elements of lower atomic number are made earliest and most easily. But the graphic zigzags up and down,

as abundance is also related to whether an element has an odd or even atomic number.

The Suess/Urey graph, together with the presence of technetium in red giants, was used to explain how all the elements could be accounted for. It was set out in a famous paper in 1957 titled 'Synthesis of the Elements in Stars', but generally known as B²FH after the names of its authors (Margaret Burbidge, Geoffrey Burbidge, William Fowler and Fred Hoyle). The paper formed the basis of stellar nucleosynthesis theory, and predicted three processes by which elements could be formed in stars and supernovae (exploding stars). It also set out the correlation between the composition of stars (determined by spectroscopy) and their age, showing that the oldest stars have the lightest elements and only newer stars contain the heavier elements.

Birth in death

A supernova event occurs when a star much more massive than our Sun reaches the end of its life. The star has fused most of its

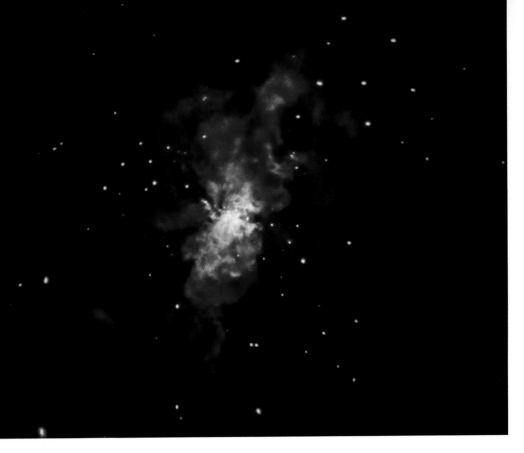

hydrogen, and more and more of its mass is pulled into its core. Or a supernova can happen when one star drags in matter from a near neighbour, taking in more than it can cope with. The more massive the star's core becomes, the greater the gravitational pull it exerts, so it pulls in yet more mass in a process that increases exponentially and

Above: Supernova SN2014J in the Messier 82 galaxy was witnessed in 2014. Eleven million light years from Earth, the supernova threw out 60 per cent of the mass of the Sun as nickel-56 (half-life six days), which decays to cobalt-56 (half-life 77 days) decaying into iron-56, which is stable.

> 'Once a star has built an iron core, there is no way it can generate energy by fusion. The star, radiating energy at a prodigious rate, becomes like a teenager with a credit card. Using resources much faster than can be replenished, it is perched on the edge of disaster.'
>
> Robert Kirshner, Harvard-Smithsonian Center for Astrophysics

ends in catastrophe. The density of the core becomes so extreme as the star collapses under its own gravity that it eventually blasts itself apart in a massive nuclear explosion which lasts several weeks.

In this brief period, the star can emit more energy than the Sun will produce in its entire life of 10 billion years or so. In the process, the dying star makes heavy elements which can't be forged in the core but can be made in the super-intense blast of the explosion. Atoms thrown out of the star are bombarded with neutrons, which enter the nucleus and split into protons and electrons to make new, heavier atoms. At last, the goal of the alchemists is reached and iron turns into gold. (And, less poetically, gold turns into lead.) These new, heavy atoms are hurled out into space where they float around in the interstellar medium until pulled into a new star forming millions or billions of years later. So they get to be part of a planetary system, and perhaps, ultimately, a life form.

We are, as Sagan says, stardust.

The question of the origins of the elements had largely been answered by 1957, and required only relatively minor tweaks over the following years. The metallicity of stars (the proportion of them that is not hydrogen or helium) increases with age as, when a star gets to the end of its life, heavier elements are forged within it. At the end, all the elements it has generated are expelled into the interstellar medium and are incorporated into new stars that form from that medium. Other elements can be made in supernovae or other extreme celestial environments.

Boron is far from boring

There is one exception to this process. Boron shouldn't really exist – its atomic configuration is such that it can't be made in stars by stellar nucleosynthesis and it probably wasn't made in the Big Bang (primordial nucleosynthesis) either. Stars tend to destroy whatever boron they start

CREATION IN REAL TIME

Supernovae happen at the rate of about one a second throughout the whole universe, or every fifty years in the Milky Way. The last supernova visible to the naked eye was reported in 1604. In 2008, for the first time, an astronomer watched a supernova explosion in real time (give or take the 88 million years it had taken for the radiation to reach Earth). Alicia Soderberg and colleagues at Carnegie-Princeton saw the five-minute burst of X-ray radiation emitted by SN 2008D 'as it happened'.

SN1006 was a brilliant supernova witnessed in AD1006 – possibly the brightest in history. The debris cloud is now 60 light years across.

off with. Along with beryllium and lithium, boron is produced by a much more sporadic and random process called cosmic-ray nucleosynthesis. In space, cosmic rays comprising very high-energy protons are abundant. As they whizz around, they often collide with carbon and oxygen in clouds of interstellar gas. The collisions (called 'spallation') cause the carbon and oxygen atoms to disintegrate and the fragments produced include boron, beryllium and lithium. There is relatively little of these elements in the universe because their method of production is so haphazard.

Native and primordial elements

It is now possible to divide the elements on Earth into three categories: primordial, native and synthetic.

Native elements are those that are found in nature. Of the 94 or so native elements, 84 are primordial. This means they have been present in the Earth from its formation at the start of the solar system. The remaining ten are the product of radioactive decay.

Primordial elements that are either diminished in abundance or of which the original stock has disappeared completely have changed into these ten other elements.

The remaining 24 elements, the synthetic elements, have been synthesized by humans and do not, as far as we know, occur naturally on Earth (though they might occur elsewhere).

Radiation keeps going

The primordial elements include four that are radioactive with a very long half-life: bismuth, thorium, uranium and plutonium. The radioactivity of bismuth-209 was only confirmed in 2003. Noël Coron at the Institut d'Astrophysique Spatiale in Orsay, France, led a team which monitored 93 g of bismuth over five days and recorded 128 alpha-particle events, indicating that just

Legend:
- Big Bang
- Small stars
- Supernovae
- Cosmic rays
- Large stars
- Made in laboratory

1 H								
3 Li	4 Be							
11 Na	12 Mg							
19 K	20 Ca	21 Sc	22 Ti	23 V	24 Cr	25 Mn	26 Fe	27 Co
37 Rb	38 Sr	39 Y	40 Zr	41 Nb	42 Mo	43 Tc	44 Ru	45 Rh
55 Cs	56 Ba	57 La	72 Hf	73 Ta	74 W	75 Re	76 Os	77 Ir
87 Fr	88 Ra	89 Ac	104 Rf	105 Db	106 Sg	107 Bh	108 Hs	109 Mt

58 Ce	59 Pr	60 Nd	61 Pm	62 Sm	63 Eu	64 Gd
90 Th	91 Pa	92 U	93 Np	94 Pu	95 Am	96 Cm

128 atoms of bismuth had decayed in that time. Its calculated half-life is 1.9×10^{19} years.

Three main decay chains occur naturally on Earth. Each one starts with a long-lived isotope: the first is thorium-232, the second is uranium-238 and the third is uranium-235. The fact that these isotopes have very long half-lives (respectively 14.05 billion years, 4.468 billion years, and 'only' 703.8 million years) explains why the chains are still going A fourth natural chain, starting with neptunium-237, is considered extinct in nature, although it has been artificially revived. Neptunium-237 has a half-life of only 2.14 million years, so there has been plenty of time since the formation of the solar system for any and all (or virtually all) neptunium to have decayed. The neptunium series has no products

with a very long half-life until it gets to bismuth-209. The sequence is shown in the table below:

Neptunium-237	2.14 million years
Palladium-233	27 days
Uranium-233	159,200 years
Thorium-229	7,304 years
Radium-225	15 days
Actinium-225	10 days
Francium-221	5 minutes
Astatine-217	32 seconds
Bismuth-213	46 minutes
Polonium-213 or Thallium-209	3.72 microseconds or 2.2 minutes
Lead-209	3.25 minutes
Bismuth-209	1.9×10^{19} years
Thallium-205	Stable

Most of the original neptunium will by now be stuck at the bismuth-209 stage, and it probably won't get out of that stage before the universe ends.

									2 He
			5 B	6 C	7 N	8 O	9 F		10 Ne
			13 Al	14 Si	15 P	16 S	17 Cl		18 Ar
28 Ni	29 Cu	30 Zn	31 Ga	32 Ge	33 As	34 Se	35 Br		36 Kr
46 Pd	47 Ag	48 Cd	49 In	50 Sn	51 Sb	52 Te	53 I		54 Xe
78 Pt	79 Au	80 Hg	81 Tl	82 Pb	83 Bi	84 Po	85 At		86 Rn
110 Ds	111 Rg	112 Cn		114 Fl		116 Lv			118 Og
65 Tb	66 Dy	67 Ho	68 Er	69 Tm	70 Yb	71 Lu			
97 Bk	98 Cf	99 Es	100 Fm	101 Md	102 No	103 Lr			

Left: The Periodic Table showing the origins of the elements.

ALL AND NOTHING

More than 2,000 years ago, people believed that all matter was created out of a void or chaos, often by a divine entity. It then arranged itself – or was arranged – from something formless into elements that bestowed qualities. In a way we have come full circle. In the modern model, all matter (space–time) came into being at the moment of the Big Bang. Over billions of years it has been forged into the chemical elements that make up and bestow qualities on everything around us. It's a model that would not appear entirely alien to the Ancient Greeks.

In uncovering the ingredients of matter and arranging them into the Periodic Table of elements, chemists have achieved an astonishing act of organization, but also one of discovery. Our categorization of living things is based on the features we have chosen to focus on and see as important. It is one of many possible categorizations. The same is not true of the Periodic Table, which is based on a fundamental aspect of matter – the number of subatomic particles in an atom. Impressively, scientists created most of the table before knowing how the sequence related to that fundamental feature, through examining its manifest effects.

The standard presentation of the Periodic Table is instantly recognizable

and iconic. There have been many attempts to present the same information differently, including circular and three-dimensional representations.

Speaking to the universe

The Periodic Table is perhaps the only piece of knowledge we might share with intelligent beings on other planets, even in other galaxies. It is valid throughout the universe. In the early 1970s, NASA launched two Pioneer spacecraft to explore the outer planets of the solar system. They are now heading into interstellar space, having left our solar system forever. Pioneer 10 is

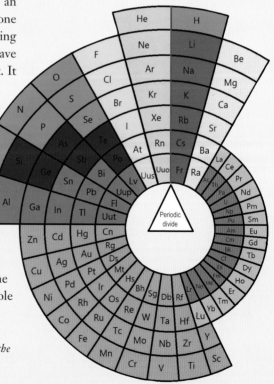

A circular Periodic Table offers a different way of visualizing the same information as presented in the familiar form.

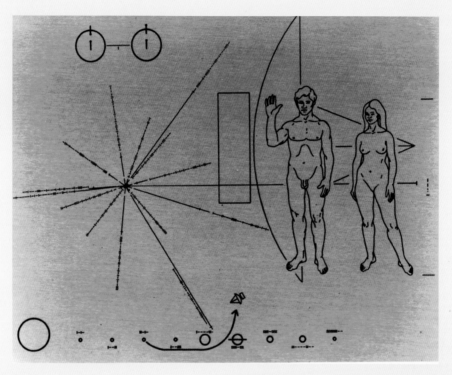

The time it takes for the hydrogen electron to flip and the wavelength of radiation emitted when it does are universal features of matter, accessible to any being in the universe able to look for it.

travelling towards the red star Aldebaran, which it will reach in two million years' time. Pioneer 11 is heading in the direction of the constellation of Aquila (The Eagle) and will pass by in four million years' time.

Both carry gold plaques with a series of symbols and pictures to tell any aliens who find them where the spacecraft came from. The first (to our way of reading) symbol on the plaque shows the flipping electron spin of neutral hydrogen, the most prolific element in the universe. It's likely that any alien civilization sufficiently advanced to trap a wandering spacecraft will be familiar with this. Elsewhere on the plaque, the units for measures of both time and distance

are related to the hydrogen atom. Time is measured in multiples of 0.7 nanoseconds, the time it takes for the spin of the hydrogen electron to flip. Distance is measured in multiples of 21 cm (8.2 in), the wavelength of the radiation emitted. (The height of the woman is eight times 21 cm.)

So universal is the language of the Periodic Table that we might use it to speak to other beings with completely different bodies, cultures and minds. What they will have discovered about the elements will be the same as we have discovered. Though the narrative might be different, the story of the Periodic Table must reach the same conclusions throughout the universe.

INDEX

Picture credits

Front cover: all images Shutterstock, apart from bottom left (Getty)

Back cover and spine: all images Shutterstock

Alamy Stock Photo: 61, 74

Archivo Municipal de Sevilla: 85 (Pepe Morón)

Bridgeman Images: 164

Diomedia: 81 (Universal Images Group/Universal History Archive), 86 (SuperStock RM/Buyen), 98–9 (Wellcome Images), 100 (Fine Art Images), 103 (Science Source), 109 (Leemage)

Mary Evans Picture Library: 39

Nicolle R. Fuller/US National Science Foundation: 188–9

Getty Images: 26–7, 41 (Gamma-Rapho), 57, 60 (UIG), 92t (ullstein bild), 105 (SSPL/Science Museum), 107 (Print Collector), 113 (De Agostini), 114–15 (Print Collector), 118 (SSPL/Science Museum), 121 (UIG), 130 (UIG), 141, 144 (SSPL/Science Museum), 158–9 (Corbis Historical), 160 (ullstein bild), 163 (AFP), 165 (SSPL), 173 (Corbis Historical), 175 (AFP), 178 (Corbis Historical), 179 (SSPL/Science Museum), 180 (Corbis Historical), 192 (National Geographic), 193 (Bettmann Archive), 196 (David Levenson)

NASA: 127 (John Hopkins University Applied Physics Laboratory/Carnegie Institution of Washington), 131, 150, 176, 187, 194 (WMAP Science Team), 198, 199

Science & Society Picture Library: 88 (Science Museum)

Science Photo Library: 10–11 (Sheila Terry), 67 (James Holmes/Hays Chemicals), 71 (Sheila Terry), 78 (Edward Kinsman), 101 (Museum of the History of Science/Oxford University Images), 116 (Ed Degginger), 119 (Emilio Segre Visual Archives/American Institute of Physics), 125, 126 (Carlos Clarivan), 128 (Science Source), 136–7, 147, 151, 169t (American Institute of Physics), 184–5

Shutterstock: 2, 6, 12, 16, 17, 28, 29, 36, 49, 54, 65, 76, 79b, 80, 83, 84, 87, 90, 91, 96, 97, 129, 132, 138–9, 145, 146, 149b, 154b, 156–7, 161, 162, 166–7, 170, 171, 172 (CERN), 181, 190

Shutterstock Editorial: 8–9 (Granger/Rex), 21 (Universal History Archive/Universal Images Group/Rex), 23t (Granger/Rex), 30–31 (Granger/Rex), 32 (Gianni Dagli Orti/Rex), 33 (Gianni Dagli Orti/Rex), 35 (Stuart Forster/Rex), 51 (Granger/Rex)

Tate Britain: 40

Wellcome Library, London: 15, 20, 22, 23b, 24, 25, 37, 38, 42, 45, 48, 50, 55, 59, 63, 72–3, 75, 77, 79t, 104, 106, 110, 111, 135, 142, 143b, 152

David Woodroffe: artworks on pages 13, 62, 66, 68, 89, 92b, 93, 123, 143t, 148, 153, 154t, 155, 169b, 174, 177, 183t and 200